# WHEELS UP

## AIRLINE BUSINESS PLAN DEVELOPMENT

### Dr. John G. Wensveen
*Embry-Riddle Aeronautical University*

THOMSON
™
BROOKS/COLE

Australia • Canada • Mexico • Singapore • Spain • United Kingdom • United States

Acquisitions Editor: Keith Dodson
Assistant Editor: Carol Ann Benedict
Editorial Assistant: Melissa Newt
Technology Project Manager: Sam Subity
Marketing Manager: Kelly McAllister
Advertising Project Manager: Brian Chaffee

Project Manager, Editorial Production: Belinda Krohmer
Print/Media Buyer: Doreen Suruki
Permissions Editor: Kiely Sexton
Cover Designer: Laurie Albrecht
Cover Printer: Transcontinental Louiseville
Printer: Transcontinental Louiseville

For more information about our products, contact us at:
Thomson Learning Academic Resource Center
1-800-423-0563
For permission to use material from this text, contact us by:
Phone: 1-800-730-2214
Fax: 1-800-730-2215
Web:
http://www.thomsonrights.com

Brooks/Cole—Thomson Learning
10 Davis Drive
Belmont, CA 94002
USA

Asia
Thomson Learning
5 Shenton Way #01-01
UIC Building
Singapore 068808

Australia/New Zealand
Thomson Learning
102 Dodds Street
Southbank, Victoria 3006
Australia

Canada
Nelson
1120 Birchmount Road
Toronto, Ontario M1K 5G4
Canada

Europe/Middle East/Africa
Thomson Learning
High Holborn House
50/51 Bedford Row
London WC1R 4LR
United Kingdom

Library of Congress Control Number:
2003115492

ISBN 0-534-39354-3

*To John,*

Only as far as we seek can we go. Only as much as we dream, can we be. I also feel quite strongly, that the things you take with us are the things you leave behind. In other words, if you try to just acquire things in your life, you can't take them with you. I seek to try to learn from others and understand. I love the power of the written word, and I like to try to share with words. When St. Exupery was flying the mail in South America one night, a young pilot didn't want to fly with him. The boss asked why, and the young pilot said, "because when St. Exupery flies, he likes to have experiences; when I fly, I don't want experiences". In a small way, I am seeking experiences on this journey. They help you see life with a greater perspective. The airplane is a tool and instrument. It is an engineering marvel. But as you glide through the air, there is a point where the science becomes an art. This is what I love most.

*Tom Claytor, somewhere in the Asian jungle!*

Tom Claytor, sponsored by National Geographic, is a bush pilot attempting to make his way around the world via all 7 continents in a single engine airplane. Tom has been a great encouragement and inspiration to me in pursuing my dreams as a scholar and business professional.

# About the Author

Dr. John G. Wensveen, born and raised in Vancouver, Canada, holds a Ph.D. in International Air Transport from the University of Wales Cardiff (United Kingdom) and a B.A. in Geography and Transportation Land Use Planning from the University of Victoria (Canada). He is currently a Professor of Airline Management with the College of Business at Embry-Riddle Aeronautical University in Daytona Beach, Florida, instructing airline management and airline operations courses at the undergraduate, graduate, and executive levels.

He is also a faculty member of the International Air Transport Association (IATA) and instructs airline and strategic management courses at IATA training centers around the world. Dr. Wensveen is an expert in airline management and concentrates on the development of business and strategic plans for air carriers of all types and sizes ranging from low-cost passenger airlines to air cargo carriers.

His professional interests include: new-entrant low-cost airlines; airline operations; aircraft fleet development; route networks; strategic cost-cutting measures; development of in-flight services; airline mergers and acquisitions; airline safety and security; airline marketing; corporate culture; corporate affairs; media relations; corporate aviation; and air cargo. Dr. Wensveen is a member of a number of professional organizations and is a frequent speaker at various international aviation events. He publishes frequently in major aviation publications and is an aviation analyst for the media often seen on major television networks and quoted in major newspapers and magazines around the world.

Dr. Wensveen is co-author of a leading aviation textbook published by Thomson Learning / Brooks-Cole entitled, *Air Transportation: A Management Perspective, 5th edition*. Dr. Wensveen has professional work experience in the Canadian, U.S. and European airline industries. He is the original founder of Orange Airways, LLC. Dr. Wensveen also has a media background and hosted his own European-based radio show for three years. He is an active consultant to airlines, airports, governments, corporate flight departments, academic institutions, and professional organizations. Dr. Wensveen has his own aviation consulting business specializing in airline management/operations, airport management/operations, aircraft acquisition, aviation training and development, and consulting to the television and film industries concerning aviation matters.

Dr. Wensveen is available for consultation concerning business plan development and travels the world on the speaker circuit talking about airline business plan development. For additional information, please contact at the following:

Dr. John Wensveen
College of Business
Embry-Riddle Aeronautical University
600 S. Clyde Morris Blvd.
Daytona Beach, FL
U.S.A.
32114
Tel: (+1)(386) 226-6700
Fax: (+1)(386) 226-6696
E-mail: wensveej@erau.edu

# Contents

# Preface

The first edition of this book on airline business plan development is aimed at a variety of readers including: academic students, both undergraduate and graduate), business professionals, and entrepreneurs. The book concentrates on business plan development suitable for airlines of all sizes, from single-engine single pilot operations to international jet operations. Regardless of a company's size, the main elements of the airline business plan remain the same.

*Wheels Up: Airline Business Plan Development*, is unique because it concentrates specifically on the airline business rather than "generic" businesses. Features other business plan books neglect, are incorporated into this book. This is more than a business plan book because modern trends are incorporated and discussed in detail to help the reader understand the importance of creating flexibility within the business plan. Flexibility is needed if an airline plans to be successful in today's increasingly competitive environment. Failure to understand recent and future trends in a new aviation environment could lead to failure.

This reference maintains various pedagogical features aimed to enhance the participant's learning experience. This includes: chapter checklists, logical organization of chapters, key terms, review questions, recommended readings, important web sites at the end of each chapter, and PowerPoint™ slides for each chapter. To enhance the reader's knowledge on airline business plan development, the author of the book created a sample business plan based on a theoretical passenger airline concept. The business plan follows a similar pattern to the chapter outline of the book. This book benefits students enrolled in college or university level air transportation programs including undergraduate, graduate and executive levels, aviation and/or management programs (MBA and Ph.D. programs), practicing air transportation professionals (airline managers, airport managers, consultants, government policy makers, investors, financiers), and entrepreneurs.

This book provides topics of special interest to readers. Each chapter provides valuable information in terms of current issues and future trends. Unfortunately, the air transportation industry changes so quickly and it is difficult to obtain published references that are up to date. The author of this book designed each chapter so that information remains as current as possible. Since the terrorist attacks on the United States on September 11th, 2001, and continued hits like the Iraqi crisis and SARS in 2003, the air transportation industry has faced many changes and challenges resulting in a new aviation environment. Readers of this book will benefit greatly by being introduced to such changes and challenges leading the reader to think about strategic decisions that will need to be made in the future.

There is a need within the air transportation field for a "one stop shop" for information on airline business plan development. This book delivers on that need and attempts to provide an understanding of how to make an idea a reality using a step-by-step process for creating a successful airline business plan for the 21$^{st}$ century. This book is aimed not only at the academic sector but also the business sector.

There are several deficiencies that exist in books in the field of air transportation and this source overcomes such deficiencies by providing a comprehensive overview of airline business plan development under one roof, acting as a handy reference to the reader, providing information on historical, current, and future trends, providing knowledge on key management strategies for the future, and providing information on the structural organization of the modern airline.

## ORGANIZATION OF THE BOOK

**Chapter 1. Introduction to the Airline Industry.** This chapter introduces the reader to the airline industry with a specific focus on global issues. The reader will gain knowledge in the following subject areas: historical trends, modern trends, and forecasts for the future. The reader will gain insight into the main issues airlines deal with and learn how unstable and unpredictable the aviation industry is. The main purpose of this chapter is to provide a "crash course" on the airline industry and prepare the reader to think about specific issues as one learns to design a business plan for an airline of any size.

**Chapter 2. The Generic Business Plan Versus the Airline Business Plan.** This chapter introduces the reader to the subject of business plan development and defines what the main purpose of a business plan is. The generic business plan is introduced and briefly discusses: Executive Summary, Description of the Business and Industry, Market Analysis, Competitor Analysis, Strategic Plan, Organization and Management Plan, Financial Plan and Financial Request, and Strategic Action Plans. The second portion of this chapter introduces the reader to the main elements of an airline business plan: Non-Disclosure Statement, Executive Summary, Business Introduction, Mission, Strategy, Market Opportunity, Analysis of Market Demand Levels, Proposed Route Structure and Schedule, Financial Analysis, Sales and Promotion Strategy, Aircraft Operating Strategy, Competition and Competitive

Response, Management and Support Team, Risk Factors, Invitation to Participate/ Finding Investment, and Certification.

**Chapter 3. Common Business Plan Mistakes.** This chapter will introduce the reader to the most common mistakes made in the development of an airline business plan. Theory will be discussed alongside real examples.

**Chapter 4. The Non-Disclosure Statement.** This chapter will introduce the reader to the non-disclosure statement by defining it and explaining the legal importance of this document. Examples of such documents will be included in this chapter.

**Chapter 5. Executive Summary.** This chapter will introduce the reader to the executive summary by defining it, explaining the main elements of this document, the "do's and Dont's", followed by a practical sample.

**Chapter 6. Business Introduction.** This chapter introduces the reader to the main elements of the business introduction including: name and type of company, passenger target market, destination(s) served, office(s) location, source of financial backing, and current position of company.

**Chapter 7. Mission.** This chapter introduces the reader to the mission statement and explains the importance of such a statement. "Do's and Dont's" are also discussed.

**Chapter 8. Strategy.** This chapter introduces the reader to the specific key steps to success and will discuss topics like; meeting the desires of the customer base, selection of optimal aircraft fleet, base of operations, outstation markets and the route map, to name a few.

**Chapter 9. Market Opportunity.** This chapter introduces the reader to discovering potential market opportunities. Research methods are explained to determine future success and possible competition.

**Chapter 10. Analysis of Market Demand Levels.** This chapter explains to the reader how to analyze market demand levels through the use of various types of data and travel characteristics, operating strategies of other airlines, and socio-economic characteristics of each potential service market. Also included is a discussion on statistical analysis methods, qualitative and quantitative research, and seasonality characteristics.

**Chapter 11. Proposed Route Structure and Schedule.** This chapter introduces the reader to designing a route structure and schedule concentrating on issues like feasibility, seasonality of demand, days and times, operational strategy, and air fares.

**Chapter 12. Financial Analysis.** This chapter introduces the reader to perhaps the most important section of the business plan. Key financial assumptions are discussed as are income statement expenses, income statement revenue, operation plan, summary financial projections for 5 year plan, and statistical summary for 5 years.

**Chapter 13. Sales and Promotion Strategy.** This chapter introduces the reader to the different methods used for selling and promotion of services. The advantages and disadvantages of such methods are discussed in detail.

**Chapter 14. Aircraft Operating Strategy.** This chapter introduces the reader to the main factors involved in acquiring and operating an aircraft. The following topics are introduced: airline strategy, types and models of aircraft utilized, configuration, technical features, the flight deck, interior features, range capability, history, technical specifications (manufacturer's table), acquisition and maintenance, aircrew training, hull and liability insurance, and safety.

**Chapter 15. Competition and Competitive Response.** This chapter introduces the reader to the competitive nature of the airline industry and explains why some airlines fail and why others prosper. Strategies leading to success are introduced.

**Chapter 16. Management and Support Team.** This chapter introduces the reader to the management team and explains key personnel required and suggested qualifications.

**Chapter 17. Risk Factors.** This chapter introduces the reader to the main risks involved in airline operations and provides strategies on how to reduce such risks. Modern airline business plans should incorporate strategies to reduce risks where possible. This chapter will have a detailed discussion on contingency planning. Additional topics include industry conditions and competition, implementation of growth strategy, operating in international markets, government regulation, number of aircraft, fuel, seasonality and cyclicality, labor relations and availability, dependence on key personnel, reliance on third parties, lack of prior operating history, and lack of public market.

**Chapter 18. Invitation to Participate (Finding Investment).** This chapter introduces the reader to the information demanded by potential investors. Timelines alongside capital requirements are discussed. Company certification, aircraft acquisition, hiring of staff, required infrastructure, and contractual agreements are discussed.

**Chapter 19. Certification.** This chapter introduces the reader to the certification process. A U.S. example will be discussed in detail walking the reader step-by-step through the 5 phases of certification. Time frames, potential barriers, and costs will be discussed.

**Chapter 20. Achieving Success.** This chapter summarizes the main elements of the airline business plan and encourages the reader to think about the main themes presented throughout the book.

## POWERPOINT™ PRESENTATION

*PowerPoint™ slides.* Additionally, complimentary PowerPoint™ slides are available as a supplement to this book. The PowerPoint™ presentation is available in downloadable form from the publisher's web site located at: *Aviation Online: The Brooks/Cole™ Aviation Resource Center* (http://aviation.brookscole.com).

## SAMPLE AIRLINE BUSINESS PLAN

*Sample business plan.* Supplementing the book is a sample airline business plan. The business plan was created exclusively for this book incorporating the elements of a proper airline business plan. The sample plan is for a fictitious company called *Utopia Airways*. The airline, based in "Utopia Land", serves a domestic and international route system using parameters that are common to the United States but could be applicable to other regions of the world as well. The sample business plan is available in CD-ROM format and the author highly recommends using the business plan alongside the book. The sample business plan is a good source for visual format but more importantly, is an excellent source showing how the financial details of an airline are put together. A number of Excel™ spreadsheets are included in the business plan showing in detail the following: expenses, revenue, and growth strategies. The sample business plan can be obtained directly from the author, Dr. John Wensveen, for a nominal fee. Contact details are provided at the start of the book.

# Acknowledgements

I consider myself a pioneer in the airline industry when it comes to writing a book of this nature because it has never been done before despite the desperate need. This book breaks new ground and will hopefully contribute to a better industry in the future. A number of barriers were faced during the initial stages of development but I quickly learned how to change such barriers into opportunities and create a product I am proud of. Writing this book created a challenging and rewarding opportunity and I sincerely owe a great deal of gratitude to the publisher and all members of the publishing team for believing in me and in this project. A very special thank you is owed to my editor, Carol Benedict. Thank you for sharing a similar passion, vision, and set of dreams.

Thank you to my colleagues at Embry-Riddle Aeronautical University for all your enthusiasm. An overwhelming amount of support was received from industry professionals from all parts of the world during the writing of this book. Thank you for your suggestions and comments. Thank you to my fellow Mifnetters for showing such an interest in this subject and promoting the book to all your friends within the industry.

Much appreciation is owed to my assistants, Benjamin Sapte and Pavel Hosa, for their outstanding contributions including numerous long days and nights. Thank you to my former student, Kenneth Byrnes, for helping come up with a title for this book.

Last but not least, I would like to say thank you to my family. Thank you to my loving parents, John and Sandra, and my sister, Kristy. I would never have made it this far in life without all of your advice and support. Thank you to my little girl, Bryanne, for letting DaDa read each chapter to you. You were my best critic!

# Part One

# Introduction to the Airline Industry

# Chapter 1

## Introduction to the Airline Industry

**Chapter Checklist**

- Historical trends

- Modern trends

- Future trends

- Purpose of a business plan

**Introduction**

It has been said that there are three types of businesses one should never invest money into: restaurants, baseball teams, and airlines. Entrepreneurs will always find ways to raise money for restaurants because there is the belief that people need to eat and they will come to "my" restaurant. A sports lover with money might invest into a losing proposition like a baseball team because it's a status symbol that might bring a positive return if they make it to the World Series. In the early days of aviation, airlines emerged because of the status involved in owning such a company. Making money was not the first priority. However, things have changed and investors continue to pour money into the airline industry despite being, arguably, the riskiest industry in the world. So, the question is, why do investors continue to invest money into an industry that will most likely not bring a return? The answer is rather simple and very attractive to those wanting to strike it rich. Although the airline industry is an extremely risky business, it is the one type of business that can bring positive returns like no other if it is successful.

This chapter introduces the reader to the airline industry with a specific focus on global issues. The reader will gain knowledge in the following subject areas: historical trends, modern trends, and forecasts for the future. The reader will gain insight into the main issues airlines deal with and learn how unstable and unpredictable the aviation industry is. The main purpose of this chapter is to provide a "crash course" on the airline industry and prepare the reader to think about specific issues as he/she learns to design a business plan for an airline of any size. It should be noted that portions of this book make reference to the United States of America, particularly when discussing certification issues. Despite this, the theory of the airline business plan applies to most parts of the world making this reference useful for any individual or organization wanting to know more about airline business plan development. Most of the subject matter discussed in this chapter is briefly introduced to the reader but is later discussed in detail in the remaining chapters.

**Historical Trends**

The airline industry is very cyclical and has been rather predictable since the early stages of airline development. Peak and trough periods can be identified throughout history creating fairly reliable data for forecasting purposes.

Without providing a lengthy history lesson, the airline industry began in 1914 with the creation of commercial flights between St. Petersburg and Tampa Bay, Florida, in the United States. The airline industry really took off in the 1930s with the emergence of new aircraft technology that allowed airlines to fly greater distances with an increased number of passengers on board. The Boeing 247 and Douglas DC-1 were the first real airliners that contributed to quick aircraft technology growth. The late 1930s to the late 1950s witnessed the growth of even larger aircraft and airline route networks turned global. The late 1950s, 1960s, and 1970s saw the growth of the jet age initiated by Boeing's 707 in 1958, the Boeing 727 in 1964, and the Boeing 747 in 1969. Supersonic travel was introduced to the world in the 1970s while advanced technologies and new aircraft concepts dominated the 1980s and 1990s.

Regulation of the airline industry happened very quickly around the world resulting in numerous acts that impacted airline operations. Many of these acts are still in place today. The biggest change in the airline industry took place in the United States when the Airline Deregulation Act of 1978 went into effect. This act essentially removed government control of the industry and created an environment of privatized airlines, reduced fares, and increased access to markets.

Different types of airlines emerged over time including the majors or legacy carriers, regional and commuter airlines in the 1980s, and new generation airlines in the 1980s and 1990s. New generation airlines include low-cost and no-frills low-cost carriers.

If this book was written prior to the terrorist attacks of September 11, 2001, the history section would have been much greater in length, as history helps predict the future. Unfortunately, the global airline industry entered a new era of aviation on that date impacting the future of the airline industry in terms of how it will evolve. Throughout the history of aviation, historical trends have been used to predict the future. Today, historical trends have little bearing in terms of forecasting the future growth of the airline industry because the industry is dealing with issues that were never thought possible. No airline business plan was ever created that incorporated contingency planning and newly evolving cost-cutting strategies to compensate for such events. The airline business plan of the future will for the first time address circumstances that were previously not imagined. The twenty-first century is the start of a new history for the airline industry.

## Modern Trends

On September 11, 2001, the world was shocked to hear about the biggest disaster in the history of aviation. Four commercial airline flights were hijacked simultaneously (United Airlines Flight 93, Newark to San Francisco; American Airlines Flight 77, Washington Dulles to Los Angeles; United Airlines Flight 11, Boston to Los Angeles; and American Airlines Flight 175,

Boston to Los Angeles). Flight 93 missed its intended target, believed to be the White House, and crashed into a field in Somerset, Pennsylvania, killing all 45 persons on board. Flight 77 was flown directly into the Pentagon, the citadel of world strategic military planning, killing 189 persons. Flight 11 was flown directly into the north tower of the World Trade Center in New York City, killing all 92 persons on board the aircraft. Flight 175 was flown directly into the south tower of the World Trade Center, killing all 65 persons on board. In the end, more than 3,000 people lost their lives on 9/11 as a result of the acts of fanatic terrorists.

Because of the events of 9/11, security at airports, as well as security at high-risk events outside aviation, was stepped up significantly. The global aviation business was hit hard financially and continues to recover. In October 2002, it was estimated that airlines in the United States would lose a total of $8 billion by the end of the fourth quarter for the same year. Some analysts said that these estimated losses were optimistic and that $10 billion would be a more likely figure.

Since the events of 9/11, a number of airlines around the world have declared bankruptcy with some closing their doors forever. In this new era of air transportation, air carriers have been forced to implement cost-cutting strategies in order to survive.

Many of today's airlines have acknowledged the need to restructure with the ultimate goal of staying alive. The days of seeking profitability seem to be something of the past as airlines around the world struggle to break even. When it comes to route structure, some airlines are attempting to survive by eliminating costly routes and replacing them with less costly routes. In other words, some airlines cannot make money on specific routes and are willing to replace such routes with new routes that also do not make money, but they are less costly to operate. What does this say about the current state of the airline industry? An optimistic analyst might respond by saying that the airline industry is in a trough period but the future looks bright. A pessimist might respond by saying that the airline industry is in a trough period as well, but the future does not look bright. Regardless of one's personal view, barriers and opportunities exist in the global airline industry despite periods of doom and gloom and periods of happy days. Barriers should be treated as opportunities rather than obstacles and that is exactly the kind of perspective airline management must take today in order to survive. New airlines starting out have a great advantage over existing carriers because they have no past to recover from.

Airlines, particularly the legacy carriers, are guilty of operating without a modern business plan. One reason is due to the high turnover rate of upper-level management meaning that the mission and vision constantly change. There is little stability in an airline environment. Secondly, the typical airline business plan is outdated and was created as a document with the inability to be flexible according to the changing environment. Flexibility is one of the key elements for achieving success in the airline industry. This topic is discussed in greater detail in upcoming chapters.

To be successful in today's environment, airlines should initiate the restructuring process by reevaluating the business plan. In many cases, established business plans will have to literally be thrown out and new ones should be created that incorporates flexibility and modern cost-cutting and marketing strategies. There is also a need to be opportunistic. Again, new airlines starting out have the advantage of incorporating flexible strategies from day one. That being said, what are the main modern trends that should be identified?

The goal of many existing airlines is to "shrink" in size while at the same time increasing market share through the joining of an airline alliance. In other words, many airlines are finding it beneficial from a cost perspective to reduce frequency to destinations served or eliminate specific routes entirely. As discussed later in this chapter, this is a current trend that will continue into the future. The larger existing airlines find this subject of interest while new start-ups and young airlines do not as such carriers tend not to belong to an alliance.

A modern airline business plan should develop route networks that are focused on point-to-point activities rather than hub and spoke networks, therefore substantially reducing costs. The point-to-point system permits greater use of the aircraft because the operational day can be expanded as the airports served tend to have less restrictive noise and night curfews versus the main hub airports. Airports in a point-to-point system are typically secondary airport facilities or smaller less congested airports with shorter taxiway systems located near a major city center. Also, because the airline is not relying on connectivity of passengers from multiple destinations, time zones are rarely an issue. If connections are not occurring outside of the immediate airline's network, there is no need to subscribe to a major computer reservation system (CRS), therefore reducing costs further.

Current trends show that modern fuel-efficient aircraft with a two pilot crew are more beneficial to flight operations versus older aircraft technology that is expensive to operate and maintain due to high fixed costs and often larger required flight deck crews. Existing airlines around the world have realized the benefits of utilizing newer technology despite the high cost of entry. In the long run, costs are significantly lower compared to operating an older aircraft.

However, many airlines find it difficult to acquire new technology due to lack of funds making acquisition through purchase next to impossible for many players in the game. Under normal circumstances, an airline will have to provide a non-refundable down payment worth 20 percent of the aircraft's value. This is money most airlines do not have. Therefore, leasing has become an attractive option for operators because down payments are not usually a requirement. Depending on the type of lease option, the airline simply provides one month's payment in advance reducing the stress of locating a rather large down payment. The lease option also permits the operator to keep the aircraft fleet current because the aircraft can be utilized for a specific period of time and traded in on a more modern aircraft.

In terms of modern aircraft being utilized in today's environment, aircraft operators of all sizes seek to fly twin-engine jet aircraft over turbo-props and multi-engine (more than two powerplants) jet aircraft. Aircraft manufacturers are able to supply the industry with twin-engine jet aircraft capable of serving market needs for short, medium, and long-haul operations. More importantly than ever before, the aircraft must fit the route it operates on.

Fleet commonality is also a current trend as aircraft operators realize the high costs involved of flying a mixed fleet of aircraft. Uniformity within the fleet reduces costs like fuel, labor, and maintenance. Incidentally, these are the three highest costs for an airline. It should be noted that the order of these costs fluctuates airline by airline.

In terms of on board services, today's airlines, particularly those operating domestic flights, have reduced or eliminated services that were once common. Many airlines are now practicing what is called maximum deck capacity. In other words, carriers are putting more passengers on board the aircraft than ever before. This is done through the reduction or elimination of First and Business Class seating configurations where the typical passenger has demands that exceed a Coach Class passenger. Food, bar, and entertainment are now considered luxury items, and many airlines are charging passengers for such services. In many cases, airlines earn more revenue by charging for such services than they earn on ticket sales.

## Future Trends

Global alliances will continue to expand among airlines because passengers demand travel to destinations beyond a single airline's network. Alliances are necessary to drive down the costs of airline operations.

There are three main factors influencing the development of airline alliances: marketing advantages, nationality and ownership rules, and competition. The marketing advantages of airline alliances were identified in the United States during the 1980s in a deregulated environment. Major U.S. airlines were able to survive in a competitive market through mergers and acquisitions, thus increasing the size of networks.

Nationality and ownership rules limit the power of airlines to purchase a foreign carrier, thus restricting competitive advantage over other carriers. Bilateral regulations state that airlines must be substantially owned and controlled by nationals of the state in which they are registered. The only way to get around these rules is to enter airline alliances that may incorporate code sharing, franchising, joint frequent-flier programs, combined sales outlets, and so on.

Competition plays a large role in the development of alliances because mutual agreements among airlines eliminate the need to compete with each other. For example, routes that were previously flown by two competing companies may result in reduced fares because there is no need to compete against each other once an alliance has been formed.

Formation of airline partnerships expands existing route networks through code-sharing agreements, provides new products for consumers, creates a high brand of service for business travelers, and creates global recognition for priority passengers. Concentration of airline activity will take place at hub airports, and priority will go to airports providing a flexible structure. As a result of transfer traffic, minimum connecting times will be important. Specific targets will be made to decrease aircraft and luggage delays.

Technology will continue to play an important role in the future development of air transportation. The advantage to technological advancement is that the passenger benefits in many ways. The number of passengers using the Internet as a tool to book reservations will increase dramatically over the next decade. Check-in procedures will be decreased and become less stressful and more convenient as passengers take advantage of electronic kiosks and remote check-in facilities from home, hotel, bus, or train station.

High labor and fuel expenses will continue to increase but strategies will be implemented to reduce these costs wherever possible. Increasing fuel prices will be combated through the utilization of modern, fuel-efficient aircraft. In many cases, airlines will replace human labor with technology. In other cases, labor costs will be reduced through different types of employment contracts. For example, a new or existing airline might consider employing different types of employees including job-share workers, part-time workers, full-time workers, and contract workers. Such infrastructure will help reduce the formation of unions due to fractionalization. Existing airlines will continue to have union representation while new start-ups will have little to no union involvement.

The corporate culture of airlines will begin to change as management acknowledges the new generation of employees coming into the workforce. The Generation Y employee has different needs and demands compared to today's employee, and organizations should provide a suitable environment to achieve increased productivity and efficiency. More airlines will attempt to copy the Southwest Airlines™ model of success. Some will succeed but many will fail. New start-ups will have a great advantage over existing airlines in terms of establishing a particular type of corporate culture. The right employee will be hired for the right job rather than hiring the wrong employee to be trained for a specific job.

In the future, airlines will subcontract much of their operations compared to today's environment, therefore reducing operating costs and risk. The "virtual" airline will emerge in greater numbers meaning that in some cases, most to all of the airline's operation will be handled by third parties. The time will come when a passenger boards an aircraft and has no idea that the airline he/she is flying on exists only in a virtual world. It will be possible to start an airline with little to no infrastructure required, as such infrastructure will be supplied outside of the airline itself. Such infrastructure includes: aircraft, employees, "below the wing" services, and "above the wing" services.

Just like current times, airlines around the world will have to deal with some similar factors in the future. Issues include increased competition, substantial economies of scale, growth through merger, mutual dependence between competing carriers, government financial assistance in regions of the world where regulation still exists, high technological turnover, excess capacity in terms of seats offered for sale, sensitivity to economic fluctuations, and close government regulation.

## Purpose of a Business Plan

The three main reasons airlines fail are due to undercapitalization, overexpansion, and inflexibility. A well structured business plan should identify such negative factors well in advance of any decision making process acting as a guide for management to realize when things are on track and when things are off track. The main purpose of a business plan is to act as a selling tool to raise money, develop ideas of how business should be conducted, while at the same time being able to assess the company's performance over time.

All too often in this industry, business plans are created but they use a format not suitable for airline operations. Unfortunately, in many cases, the airline is doomed from the start because of the incorrect format. *Chapter Two* introduces the reader to business plans and distinguishes the differences between a generic business plan and an airline business plan. This book is the world's first suitable reference that describes how to write a business plan for an airline of any size and is a must read for anyone interested in airline business plan development.

## Summary

The airline industry is perhaps the riskiest industry to invest money into, and yet investors are continuously attracted to the industry regardless of its positive or negative state. The main reason that the airline industry is attractive to investors is because it is one of few industries that can provide a massive payback if things work out well. This chapter introduced the reader to the global airline industry briefly discussing historical trends, current trends, and future trends. It was acknowledged that historical trends have been good indicators of how the airline industry evolved over time. However, such trends have little validity in today's environment because the airline industry is facing unprecedented challenges that will change the future of how the industry evolves forever.

The airline industry is unstable and unpredictable, forcing airlines to restructure with the ultimate goal of implementing effective cost-cutting strategies to achieve break-even levels and hopefully profit. For the first time in the history of the airline business, existing airlines are focusing energies on business plan development, recognizing the need to create flexible strategies that can change as the external operating environment changes. New airline start-ups have a great advantage over existing carriers in terms of implementing such strategies because they have no past to deal with in terms of recovering losses or creating a new business strategy. This chapter simply introduced the reader to the global airline industry briefly touching on the main factors the industry faces. The remaining chapters concentrate on all the factors mentioned in this chapter in greater detail along with the introduction of additional information.

**Key Terms**

Peak period

Trough period

Major carrier

Legacy carrier

Point-to-point network

Hub and spoke network

Secondary airport

Computer Reservation System (CRS)

Fleet commonality

Maximum deck capacity

Global alliance

Undercapitalization

Overexpansion

Flexibility

**Review Questions**

1. What role did historical trends play in the airline industry leading up to the terrorist attacks of September 11, 2001? How did the role of historical trends change after the tragic events?

2. For the first time in the history of aviation, airlines have acknowledged the need to restructure. What does restructuring encompass in terms of current and future trends, and how does the restructuring process impact the business plan?

3.  The airline business plan has a number of objectives. What are those objectives, and what key ingredients should be incorporated in order to achieve those objectives?


**Web Sites**

http://www.faa.org

http://www.airwise.com

http://www.flyaow.com

http://www.aviation-history.com

http://www.thehistorynet.com/AviationHistory


**Recommended Reading**

Abeyratne, Ruwantissa. *Aviation Trends in the New Millennium.* Burlington, VT: Ashgate Publishing, 2004.

Dempsey, Paul Stephen and Gesell, Laurence. *Airline Management: Strategies for the 21$^{st}$ Century.* Chandler, AZ: Coast Aire Publications, 1997.

Graham, Brian. *Geography and Air Transport.* Chichester, UK: John Wiley & Sons, 1995.

Holloway, Stephen. *Straight and Level: Practical Airline Economics.* Burlington, VT, and Aldershot, UK: Ashgate Publishing, 1993.

Morrell, Peter. *Airline Finance.* Brookfield, VT: Ashgate Publishing, 1997.

Radnoti, George. *Profit Strategies for Air Transportation.* New York, NY: McGraw Hill, 2002.

Shaw, Stephen. *Airline Marketing and Management, 4$^{th}$ Ed..* Burlington, VT: Ashgate Publishing, 1999.

Tretheway, Michael and Oum, Tae. *Airline Economics: Foundations for Strategy and Planning.* Vancouver, Canada: University of British Columbia, 1992.

Wells, Alexander T., and Wensveen, John G. *Air Transportation: A Management Perspective, 5th ed..* Belmont, CA: Thomson-Brooks/Cole™, 2004.

# Chapter 2

## The Generic vs. Airline Business Plan

**Chapter Checklist**

- Elements of a generic business plan

- Elements of an airline business plan

**Introduction**

*If you have no plan, you plan to fail!*

Creating a business plan is a complex and tedious task but it is important to have such a plan if a company is to be successful. All too often, small companies go into business without a business plan and eventually find themselves in trouble because there is no formal structure to achieve success. A well thought out business plan guides the company and functions as a map to achieve positive results. The document can be consulted to see if the company is deviating from its ultimate goal and it can be used to guide its future direction. Without such a plan of action, it is likely that the company will fail. That being said, most large organizations have a business plan but that doesn't mean the company is guaranteed success. This is especially true in the airline business. Airlines of all sizes neglect to insert one key ingredient when it comes to their business plan structure – flexibility!

Many experts agree that airlines fail because of two key factors: overexpansion and undercapitalization. While this is true, there is a third factor that must be included: inflexible business plans. Historically, many airlines failed because of this third factor. Not being able to adjust the business according to the changing environment in which the airline operates, strangles the company to the point where it becomes dependent on a life support system, usually in the form of government subsidies, loans, or employee share plans. In short, not being flexible prolongs the life of the company until it can no longer breathe resulting in bankruptcy, a merger with a competing carrier, or permanent closure. The terrorist activities against the United States on September 11th, 2001, the Iraqi crisis and the SARS crisis of 2003, changed the world forever, including the way airlines of all sizes operate. The global airline industry is operating in a period of survival, adaptation, and recovery resulting in the need for flexible business strategies more than ever.

This chapter introduces the reader to the elements of a generic business plan followed by a discussion on the elements of an airline business plan. Although both types of business plans have the same function, they are very different in terms of how they evolve. Most airlines of the past, and today, use the elements of a generic business plan not realizing that their future could be doomed from the start.

This chapter introduces the reader to the subject of business plan development by distinguishing the main differences between a generic business plan and an airline business plan.

**Elements of a Generic Business Plan**

A business plan must contain certain elements or key factors in order to be effective. Unfortunately, many people do not understand the concept of a business plan and, therefore, leave out some of these elements which leads to a potentially disastrous outcome. *Chapter One* discussed the purpose of a business plan and described it as a tool to raise money, develop ideas of how the business should be conducted, while at the same time being able to assess the company's performance over time. Although there is no single template or map to follow when it comes to writing a business plan, most types of businesses will be designed using the elements of a "generic" business plan. In many cases, these elements are enough to satisfy the requirements of a business plan but when applied to the airline industry, these elements are not sufficient because they neglect to concentrate on additional elements that must be focused on. The design of a business plan differs by industry and by type of company and, therefore, must be customized accordingly. *Figure 2.1* shows the structure of a generic business plan as it might apply to any type of business.

A lot of people make the mistake of purchasing business plan books, business plan software, or downloading free samples of business plans from the Internet thinking that the information they acquired is going to lead to the formation of a successful business plan. Sellers of such products know the demand for these products and capitalize on it by selling generic outlines regardless of what type of industry the information will be applied to. It is amazing to see how many "one-stop shops" are on the market for buying such generic information. Purchasers of such products should be very cautious because the information they acquire is not necessarily accurate.

There are also many consultants offering their services and "expertise" to design business plans for anyone seeking it. Again, purchasers should be very cautious when dealing with an organization that says it can design a business plan for any type of industry. Would you go to an Italian restaurant for Mexican food? If such a document is to be designed by a consultant, make sure that the person or organization you hire is a credible expert in the field. In the aviation industry, there are many individuals offering such services but there are a limited number of "qualified" experts. A lot of consultants will simply use an existing template and fill in the blanks so to speak while charging the customer outrageous sums of money. Often, such templates were created for other airline concepts that failed.

**Fig. 2.1**
Elements of a Generic Business Plan

Executive Summary
Non-Disclosure Statement
Description of the Business and Industry
Market Analysis
Competitor Analysis
Strategic Plan
Organization and Management Plan
Financial Plan and Financial Request
Strategic Action Plans

**Elements of an Airline Business Plan**

As previously discussed, although all business plans have the same purpose, there is no such thing as one type of business plan. Business plans differ by industry and by type of company. There is no recognized single source that one can go to in order to determine what the main elements of an airline business plan are. This book is the first of its kind and offers to the reader a "blueprint" on how to construct a proper airline business plan. *Figure 2.2* outlines the main elements of an airline business plan in the order that each element should be addressed in the actual business plan. Each of these elements is discussed independently in its own chapter creating the main focus of this book.

**Fig. 2.2**
Elements of an Airline Business Plan

Executive Summary
Non-Disclosure Statement
Business Introduction
Mission
Strategy
Market Opportunity
Analysis of Market Demand Levels
Proposed Route Structure and Schedule
Financial Analysis (statements from business plan)
Sales and Promotion Strategy
Aircraft Operating Strategy
Competition and Competitive Response
Management and Support Team

Risk Factors
Invitation to Participate

The above structure of an airline business plan can be applied to any size of business regardless of its nature ranging from a single-engine propeller driven aircraft to an international operation utilizing a fleet of long-range jet aircraft. Following this step-by-step guide will lead to the creation of a successful plan. This structure is useful to a new organization starting out as well as an existing organization seeking to restructure its business plan. Since the events of 9/11, the term restructuring has become a part of virtually every airline's vocabulary. The main focus of any restructuring plan should be on the redevelopment of the business plan. Any airline not following the format in *Figure 2.2* should seriously consider following this format. New air carriers starting out have a major advantage over existing carriers in terms of their potential success if the above mentioned structure is implemented from day one.  The generic business plan is no longer valid in the new era of aviation. As the saying goes, "it's time to do away with the old and get with the new".

## Summary

The purpose of a business plan is to act as a selling tool to raise money and to be used as a benchmark or a map of how the company is performing. This chapter introduced the reader to the main differences between a generic business plan and an airline business plan. Although both types of business plans have the same function, they are very much different in terms of format meaning the difference between success and failure for the airline. Historically, airlines have operated using a generic business plan structure with some modifications. For existing airlines, it has been found that the generic format has resulted in complications for the company mainly in the form of not having the ability to be flexible as the operating environment changes. The elements of the airline business plan are tailored to the aviation industry and if written correctly, such a business plan permits flexibility. In order to be successful in a rapidly changing aviation environment, it is important for new companies to be designed using the elements of the airline business plan. For existing companies contemplating restructuring operations, step one should be the implementation of a new business plan based on the elements of an airline business plan. Such a decision could save thousands or even millions of dollars in the long term.

## Key Terms

Business plan

Overexpansion

Undercapitalization

Flexibility

Generic business plan

Airline business plan

Restructuring

## Review Questions

1. What are the main functions of a business plan?

2. What are the differences between a generic business plan and an airline business plan?

3. Does the business plan allow for flexibility in a rapidly changing aviation environment? In other words, is the business going to be impacted as the operating environment changes and does the business plan permit changes or contingency planning?

4. As the airline grows, is the business plan structured in a way that it does not have to be seriously altered? If changes have to be made to the business plan, how will these changes impact the success or failure of the airline?

## Web Sites

http://www.business-plan-success.com

http:///www.bizplanit.com

http:///www.businessvision.co.uk

**Recommended Reading**

Shaw, Stephen. *Airline Marketing and Management, 4$^{th}$ Ed..* Burlington, VT: Ashgate Publishing, 1999.

Tretheway, Michael and Oum, Tae. *Airline Economics: Foundations for Strategy and Planning.* Vancouver, Canada: University of British Columbia, 1992.

Wells, Alexander T., and Wensveen, John G. *Air Transportation: A Management Perspective, 5$^{th}$ ed..* Belmont, CA: Thomson-Brooks/Cole™ , 2004.

# Chapter 3
## Common Business Plan Mistakes

**Chapter Checklist**

- Airline business plan mistakes

- Myths to avoid

## Introduction

All too often, developers of airline business plans make many mistakes starting from the initial stages. In order to write a business plan, there must be an "idea". The idea stage is usually the first phase of development and it permits exploration of the business concept. As the concept is explored in greater depth, barriers or obstacles are found as well as opportunities. Historically, many airline business plans never got off the ground or failed once they did because the original idea became the center of the business plan. In other words, the creator(s) did not recognize potential barriers or neglected to act upon them. Most of the time, the original idea will not be the idea that the business plan is designed upon. That being said, it is important to incorporate flexibility into the airline business plan and do not create a "fantasy plan". Be realistic, conservative, and aim for steady moderate growth.

This chapter introduces the reader to the most common mistakes made in the development of an airline business plan, including mistakes made in all the phases of development.

## Business Plan Mistakes

It is important to note that no business plan is perfect and changes will occur once the airline is operating which will lead to creating deviations from the original business plan. However, it is possible to create a near perfect business plan if common mistakes are known in advance. This section outlines five key areas where mistakes are frequently made.

## Capturing the Reader's Interest

The final version of the business plan will be read by numerous individuals for the purposes of raising capital, establishing contracts, and obtaining certification. Capturing the reader's interest from the start is very important because the reader will most likely be a busy individual with many obligations. This is especially true when dealing with professionals involved in the raising of money. Typically, such people read hundreds or even thousands of business plans per day looking for the "right" one. In short, the business plan must capture the reader's interest within the first 90 seconds; otherwise, it might end up on the floor, in the garbage bin, or filed as a "reject". The airline business plan must have aesthetic appeal at first glance meaning the format must be capturing and well organized encouraging the reader to turn the pages. If the business plan does not meet such requirements, it is possible that it may never be read.

**Inaccuracies, Inconsistencies, Lack of Objectivity**

The airline business plan must be accurate and thorough at all levels as the reader will be focusing on the content once interest has been captured. A minor error could be as detrimental as a major error depending on the audience reading the document. It is very important that the reader is not distracted from the content of the airline business plan and negative bias must be avoided at all times.

Many airline business plans have failed because they were unrealistic. When raising money to launch an airline, the developer(s) of the business plan often attempts to make the business concept look too good believing that any negativity would cause the idea to fail before it ever got off the ground. A smart investor knows the difference between fantasy and reality, therefore, the business plan should never try to "sugar coat" reality. The concept stands little chance of coming to life if the developer(s) does not produce an honest business plan. Always produce a realistic picture of the business opportunity. Examples of being unrealistic include: not being practical and pragmatic, underestimating time and amount of money required, overestimating potential, underestimating the competition, inflated numbers related to financials, and inflating expectations.

Clarity is extremely important in the design of the airline business plan and although the concept might be good, it can go unnoticed because of this factor. The airline business plan contains an overwhelming amount of detail and must be structured in a clear and concise format. In today's financial world, a shorter business plan is often more important than a lengthy one. Examples of mistakes often associated with clarity are: unclear and overoptimistic, too detailed, too much useless information, too many numbers, not able to present the need for funding in a simple manner, and difficulty explaining the real business.

Incompleteness is a common mistake made in the development of the airline business plan. How is this possible given the amount of information provided in the document? The answer is simple. It is associated with the lack of background work or homework into the market and the competition. Too many airline business plans make assumptions rather than presenting the actual facts. Historically, the major airlines never asked the customer what their actual needs and demands were. Instead, the airlines assumed what those needs and demands were. In today's environment, airlines are going to great efforts to ask the customers what they want from an airline. Finally, the airlines have realized there is little passenger loyalty in a very competitive industry. Knowing the needs and demands of the customer over the competition is powerful information in terms of attracting passengers to the airline. A complete airline business plan will demonstrate thorough knowledge of the market resulting in increased credibility. Examples of mistakes associated with incompleteness include: not showing profit timeline, poor presentation, lack of defined objectives, poor executive summary, and insufficient explanation of marketing and sales strategy.

*Sustainable, Competitive Advantage*

Establishing sustainable, competitive advantage in the airline business plan is difficult but it must be done. When looking at the business plan, the reader is primarily interested in knowing what the market is. Does the market actually exist? How will this company fit into that market? Does this airline have the ability to compete with other carriers and if so, what will be the long-term gain? It is very important not to underestimate the competitors. If no competition exists, never assume it will not happen. Always assume that competition will happen and build a contingency plan on how the airline will survive not if it happens but when it happens. The airline business plan should identify what the company's strengths are and how these strengths will be used to gain competitive advantage.

Failure to identify market need is a common mistake made in the development of the airline business plan. The business plan must be superior to any potential competitor's business plan and should be able to gain the investor's attention and money. Examples of mistakes associated with market need include: inadequate presentation of market need, failure to identify the size of the market, failure to identify the niche market the airline will operate in, and failure to identify why the airline will be superior to any other. In other words, what is the difference between this airline and its competition? What sets this airline apart from the others? Adding uniqueness to the business concept is important otherwise, the airline is like any other and will most likely fail to raise the funds needed to fly.

*The Management Team*

Choosing the right person for the right job is important in the management team selection process. In many airline business plans, the importance of the management team is often underestimated. Many investors believe that a superior management team can make a mediocre idea successful. This was proven in the case of Continental Airlines when the airline filed for bankruptcy in 1990. A new CEO, Gordon Bethune, was put into power and the management team was restructured. A failing airline was turned around into one of the most successful carriers under a new management team. A strong management team will help reduce any doubts the investor might have about the ability of the airline to be successful. The airline business plan must promote the team and introduce key advisors that will assist in the company's success. The management team becomes important in outlining strategic objectives and implementation of the business plan. A good team will be fully aware of all risks involved and such risks should be highlighted in the actual business plan. With each risk should appear a solution or strategy on how the airline will deal with each risk. Failure to provide a thorough SWOT analysis (Strengths, Weaknesses, Opportunities, Threats) combined with strategies will enhance the investor's doubt factor.

In many cases, the developer(s) of the airline business plan often does not believe that the management team introduced in the business plan could be a major weakness. However, investors often see this as a major weakness and can be critical in deciding whether money will be invested or not. Examples of mistakes associated with the management team include: not focusing on the management team and its experience and lack of information on management or inexperience in the field.

*Revenue Growth and Profitability*

Many airline business plans fail to demonstrate revenue growth and profitability. Historically, top line growth is mentioned but bottom line growth is not. In order to sell the business concept, bottom line growth is essential and this growth must be based on credible financial assumptions. It is important to have quantitative sections backed up by qualitative sections. All financial projections should be consistent with accepted accounting principles. Unrealistic financial projections should not be incorporated into the airline business plan. In most cases, being able to show "less" rather than "more" can be very powerful in terms of obtaining interest. The financial section of the business plan determines the success or failure of the business concept. If the numbers are falsified, the company will fail. Be realistic with numbers and follow a simple lesson. Whatever the actual costs work out to be, double them. Whatever the actual profits work out to be, half them. Implementing this philosophy will create a strong, well-rounded, realistic business plan. Examples of common mistakes made associated with revenue growth and profitability are: failure to show how ROI (return on investment) will be generated for investors, no clear ROI, lack of return on investment figures, too much concentration on financial numbers, vague assumptions regarding potential cash flow, lack of understanding business start up costs, and lack of research.

**Myths to Avoid**

"We don't need a business plan"

"We need a business plan, but not right away"

"We need a business plan, but we can do it ourselves"

"We need a business plan, but we can't spend more than $X$ dollars"

"We only have to do this once"

**Summary**

When it comes to development of the airline business plan, many common mistakes are often made despite the previous attempts to achieve success by others in the same industry. These mistakes are constantly repeated and could be avoided with proper research. Unfortunately, timing is critical when putting together a business plan and resources are often lacking. Many developers of such plans attempt to use the wrong tools and combine the tool selection with assumptions. Failure is a strong possibility right from the start. This chapter identified five common mistakes made in the development of the airline business plan followed by a brief discussion on each mistake. Knowing the common mistakes frequently made and doing the right background research will be contributing factors to creating a solid airline business plan.

**Key Terms**

Competitive advantage

SWOT analysis (Strengths, Weaknesses, Opportunities, Threats)

ROI (Return on Investment)

**Review Questions**

1.  Does the airline business plan capture the reader's interest and does it cater to all potential audiences?

2.  Is the airline business plan accurate, consistent, and objective?

3.  Does the airline business plan show that the business plan concept will gain sustainable competitive advantage? If so, how is the business concept different from the competition?

4.  Is the airline business plan supported by an experienced management team and does each member have something valuable to contribute?

5.  Does the airline business plan demonstrate revenue growth and profitability? If so, has bottom line growth been identified along with top line growth?

6.  Are financial projections realistic?

## Web Sites

http://www.business-plan-success.com

http:///www.bizplanit.com

http:///www.businessvision.co.uk

## Recommended Reading

Abeyratne, Ruwantissa. *Aviation Trends in the New Millennium.* Burlington, VT: Ashgate Publishing, 2004.

Arkebauer, James. The McGraw-Hill Guide to Writing a High-Impact Business Plan: A Proven Blueprint for First-Time Entrepreneurs, McGraw-Hill, New York, 1994

Holloway, Stephen.  *Straight and Level: Practical Airline Economics.* Burlington, VT, and Aldershot, UK: Ashgate Publishing, 1993.

Morrell, Peter. *Airline Finance.* Brookfield, VT: Ashgate Publishing, 1997.

Piotrowski, Maryann. Effective Business Writing: Strategies, Suggestions, and Examples, a Guide for Those Who Write on the Job, Harper Collins, New York, 1996

Tretheway, Michael and Oum, Tae.  *Airline Economics: Foundations for Strategy and Planning.*  Vancouver, Canada: University of British Columbia, 1992.

Wells, Alexander T., and Wensveen, John G. *Air Transportation: A Management Perspective, 5th ed.*. Belmont, CA: Thomson-Brooks/Cole™ , 2004.

# Part Two

# The Airline Business Plan

# Chapter 4
## The Non-Disclosure Agreement

**Chapter Checklist**

- Purpose of the Non-Disclosure Agreement

- Do I need a lawyer?

**Introduction**

Protection is the key when creating the concept to create an airline or any type of company for that matter. All too often, people with great ideas have their dreams taken away because they did not take the appropriate measures to protect their idea from the start. A *Non-Disclosure Agreement (NDA)*, sometimes referred to as a *Confidentiality Agreement*, does not guarantee 100% protection. However, it is perhaps the best tool one has access to in order to avoid the concept from being "stolen". Although this chapter is very brief, it is one of the most important chapters in this book because it shows the reader how to create a legally binding form for protecting the business concept. This could mean the difference between dream and reality.

**Purpose of the Non-Disclosure Statement**

Before creating the Non-Disclosure Agreement, it is advisable to take one important step prior to building the business plan. Once the idea for the air carrier is born, write down all of the ideas, good and bad, in bullet point form. Once completed, insert the ideas into a self-addressed stamped envelope (SASE) and put it into the mail. Once the envelope is received, it is important not to open the package. The sealed package combined with a postmark, provides substantial proof of the originator's idea in the event that such information is required.

The Non-Disclosure Agreement should be written once the business plan has been completed and is ready to be marketed for the purpose of raising capital or for contract negotiations. At no time, should the details of the business plan be shown to anyone unless they have signed the agreement. If an individual is not prepared to sign the agreement, it is recommended that the relationship be ceased. This rule also applies to any individual involved in the creation of the business plan. If more than one individual is involved in the creation of the company, all participants should sign the agreement. Once all participants have signed the agreement, the paperwork should be filed and kept in a secure place in the event that such information is needed.

The main purpose of the Non-Disclosure Agreement is to protect the originator of the business plan concept from having similar ideas "stolen" or exploited. The agreement can be basic or very detailed depending on what information the owner of the business plan wishes to present. *Figure 4.1* provides a good example of a basic but descriptive Non-Disclosure Agreement. A more complicated example might include the following clauses within the document: Purpose, Definitions, Use of Confidential Information, Non-Disclosure Obligations, Return of Confidential Information, Non-

Assignable, Solicitation, Governing Law, No License, Indemnity, Building on Successors, No Publicity, No Export, Term and Termination, Severability, Attorney Fees, Entire Agreement, and Arbitration.

## Do I Need a Lawyer?

A lawyer is not required in order to create a solid Non-Disclosure Agreement. However, it is recommended to seek the advice of a lawyer when preparing any type of legal documentation, especially when designing a project of this nature. Spending a few hundred dollars on a lawyer could mean the difference between earning a lot of money in the future and seeing the dream crash before it ever takes off. If you are or have visions of being an entrepreneur then you should know the difference between getting value for your money as opposed to focusing solely on absolute costs, which often disregards value. If one seeks the advice of a lawyer, be sure to locate a reputable firm specializing in corporate or contract law.

There are a number of companies advertising samples of Non-Disclosure Agreements on the Internet. These forms can be downloaded for an average cost of approximately $20.00 USD. If such samples are used, the buyer should be aware that the forms are generic in nature and might require editing to meet the needs of your business concept.

**Fig. 4.1**
Non-Disclosure Agreement

The undersigned acknowledges that INSERT NAME OF COMPANY OFFERING THE BUSINESS PLAN has furnished to the undersigned certain proprietary data ("Confidential Information") relating to the business affairs and operations of INSERT NAME OF AIRLINE for study and evaluation.

It is acknowledged by the undersigned that the information provided by INSERT NAME OF COMPANY OFFERING THE BUSINESS PLAN is confidential; therefore, the undersigned agrees not to disclose it and not to disclose that any discussions or contracts with INSERT NAME OF COMPANY OFFERING THE BUSINESS PLAN have occurred or are intended, other than as provided for in the following paragraph.

It is acknowledged by the undersigned that information to be furnished is in all respects confidential in nature, other than information which is in the public domain through other means and that any disclosure or use of same by the undersigned, except as provided in this agreement, may cause serious harm or damage to INSERT NAME OF COMPANY OFFERING THE BUSINESS PLAN, and its owners and officers. Therefore, the undersigned agrees that the undersigned will not use the information furnished for any purpose other than as stated above, and agrees that the undersigned will not either directly or indirectly by agent, employee, or representative, disclose this information, either in whole or in part, to any third party; provided, however that (a) information furnished may be disclosed only those directors, officers and employees of the undersigned and to the undersigned's advisors or their representatives who need such information for the purpose of evaluating any possible transaction (it being understood that those directors, officers, employees, advisors and representatives shall be informed by the undersigned to treat such information confidentially), and (b) any disclosure of information may be made to which INSERT NAME OF COMPANY OFFERING THE BUSINESS PLAN consents in writing. At the close of negotiations, the undersigned will return to INSERT NAME OF COMPANY OFFERING THE BUSINESS PLAN all records, reports, documents, and memoranda furnished and will not make or retain any copy thereof.

_____       _____
Signature                                                    Date

_____
Name (typed or printed)


INSERT NAME, ADDRESS, AND CONTACT DETAILS OF COMPANY OFFERING THE BUSINESS PLAN
*This is a business plan. It does not imply an offering of securities.*

## Summary

The purpose of the Non-Disclosure Agreement is to protect the creator of the business plan from having the contents of the plan stolen or manipulated by an outside party. This is what is referred to as proprietary information. There are basic and descriptive forms of such agreements and the type of agreement used is dependent on the amount of freedom the creator wishes to offer to potential readers. Generic templates for Non-Disclosure Agreements are available to purchase on the Internet but it is recommended that a lawyer be consulted when preparing any type of legal documentation related to this type of business.

## Key Terms

Non-Disclosure Agreement

Confidentiality Agreement

Proprietary Information

## Review Questions

1. Why is it important to protect the business idea and at what stage(s) is this done?

2. What is the purpose of a Non-Disclosure Agreement? What are the different types of Non-Disclosure Agreements that exist?

3. Is it beneficial to purchase a Non-Disclosure Agreement from the Internet? Is it important to seek the advice of an attorney when preparing any type of legal documentation?

4. Have I as the creator of the business plan protected the idea to the best extent possible?

## Web Sites

The following web sites are links to companies providing various samples of a non-disclosure statement. Please note, in order to use such information, each company charges the user to download such information.

http://www.LawDepot.com

http://www.confidentialityagreementkit.com/

http://www.lodgix.com/

http://www.findlegalforms.com

http://www.confidentiality.biz/

http://mytechnologylawyer.com

**Recommended Reading**

Loffel, Egon W. *Protecting Your Business*. New York, NY: D. McKay Co., 1977.

National Air Transportation Association. *Negotiating Aviation Agreements: Guidelines for Building and Protecting Your Business*, 2nd edition. Alexandria, VA: National Air Transportation Association, 1986.

Nicholas, Ted. *The Complete Guide to Business Agreements*. Chicago, IL: Enterprise Dearborn, 1992.

Stim Richard, Fishman Stephen. *Raising Venture Capital: Leading VC's Reveal What They Really Look for to Make Investments*. Boston, MA: Aspatore Books, 2003.

# Chapter 5
## Executive Summary

**Chapter Checklist**

- Executive Summary defined

- Main elements of the Executive Summary

- The "Do's" and "Dont's"

## Introduction

The Executive Summary is equal to the business plan in terms of importance and in some cases, depending on who the audience is, plays a more significant role. The Executive Summary is written once the business plan has been completed providing a summary of the business plan without giving away the answers. The objective of the Executive Summary is to create what is called a "teaser package" and to sell the business idea to a variety of interest groups. In the case of the airline industry, interested parties might include investors, airports, manufacturers and suppliers, or the government.

This chapter introduces the reader to the executive summary by defining it, explains the main elements of this document, and highlights the "do's" and "Dont's".

## Executive Summary Defined

Defining the Executive Summary is a rather simple task. However, designing such a document is a very complex process resulting in success or failure of the airline. The main objective of the Executive Summary is to draw attention to the business concept in the case of a new company starting out or to enhance interest of an existing company. In most cases, the document is used to stimulate investment to get the company off the ground or to expand it. In other cases, the Executive Summary is used as an aid in the establishment of contracts where it is important for the airline to prove its credibility. For example, a new airline starting out may use the Executive Summary to show its credibility with the airports within its route network when setting up contracts for flight operations, passenger and baggage ground handling, catering, or fueling.

The Executive Summary is written once the business plan has been completed in full and it is expected that it will be subject to no further changes. It should be noted that some changes will be made to the business plan once capital has been raised, management has been placed, strategies have been altered, or changes have occurred in the operating environment. Again, the main objective of the Executive Summary is to attract the interest of potential investors and there are only a few moments that exist to capture one's attention. The document must be clear, concise, and should highlight important points. The general business concept should be clear at a quick glance followed by the amount of money required, a cash flow analysis, and expected return on investment (ROI). It is also advisable to provide the reader with contact information for the legal and accounting firms used or expected to be used. Such information increases the validity of the business plan and shows the reader that you have done your homework.

The average audience will take two or three minutes to review the Executive Summary meaning that many important pieces of information will not be read at all. To compensate for this, wording should be kept to a minimum and bullet points should be used where necessary to emphasize a point or various points. The reader should walk away with the following information if nothing else. First, the reader should have a general understanding of what the business concept is. Second, the reader should know how much money is required. Third, the reader should have some knowledge on the expected rate of return on his/her investment. If the reader is not provided this information, then the Executive Summary has not done its job and the chances of gaining one's attention will most likely fail.

Some business experts will argue that an Executive Summary should not exceed two to four single-spaced pages in length. This is especially true when trying to obtain initial funding from a venture capital (VC) firm. However, other business experts will argue that the Executive Summary should be longer. When presenting the Executive Summary, one should assume that the reader has limited time and it is in the best interest of all concerned to be brief yet descriptive at the same time. Key questions that must be answered in the Executive Summary are presented at the end of this chapter. Attempt to answer these questions briefly but thoroughly.

## Main Elements of the Executive Summary

The Executive Summary is a mini version of the airline's business plan. Just as the business plan has certain elements, so does the Executive Summary. Each element summarizes one part of the business plan and should be designed in an orderly fashion so that flow or smooth transition is created as the reader looks at the document. The Executive Summary should include eleven main elements.

### The Company

This section of the Executive Summary should highlight the basic details of the company in terms of its location and contact details, type of company (i.e., flight school, charter, air taxi, shuttle, non-scheduled carrier, scheduled carrier), and route network. In the case of an actual airline, it is important to identify whether the company is no-frills, low-cost, or luxury oriented.

*Company's Mission*

Every organization, regardless of what type of business it is, should have a well-defined aim otherwise known as a mission statement. The mission statement should not exceed one or two sentences and should be as descriptive as possible using a limited number of words. Surprisingly, many companies of all sizes do not have a mission or overall aim. If this is the case, then achieving success can be difficult because there is no clear objective.

*Products and Services*

Is the airline business product oriented or service oriented? There is no general consensus on this issue. For the purposes of this book, the airline business is both product and service oriented. The main product is the seat being offered for rent to transport the passenger between points. In the case of air cargo, space is the product being rented. The interaction that occurs between the customer and the employee is considered service. In today's environment, many might argue there is no such thing as service in the airline business. The Executive Summary should highlight the main product(s) and services being offered based on the type of carrier (i.e., charter class, economy class, business class, first class).

*Marketing and Sales Strategy*

This section of the Executive Summary should indicate how the company will market the products and services offered. Will the company use direct selling methods like the Internet or telephone? Will travel agents or tour operators be used? Will the company utilize unique marketing methods to obtain sales? These are strictly examples of types of questions the Executive Summary should answer.

*The Competition*

Once the business plan has been completed, all competitors and potential competitors should be identified. It is important to outline in the Executive Summary that competition exists or could exist but the actual names of the competitors do not have to be identified at that time as they will be discussed in the actual business plan. If the Executive Summary says no competition exists or will ever exist, a red flag will be raised in the mind of the reader. Never assume that competition will ever exist.

*Target Market*

This section should highlight what type of passenger and/or product the business concept caters to. For example, will the airline be flying leisure-oriented passengers, business oriented passengers, or will another classification apply? Be sure to make the reader clear on what the actual target market is.

*Management*

The type of personnel operating the company is very important and the Executive Summary might indicate the name, title, and brief background of each key decision maker. It is very important to make sure that the "right" person is in the "right" position. An experienced management team is necessary to attract attention. However, to keep the Executive Summary brief, it is possible to avoid inserting the information above as long as you can convince the reader that the right management has been selected.

*Operations*

How will the day-to-day operations of the company be managed? This section of the Executive Summary should briefly outline the main base of operations followed by information on the general conduct of the business. Questions to answer might include the following: What is the role of management? Will portions of the business be subcontracted? If so, will this include portions of the business or the entire business? This section leaves a great amount of room for variety and it is completely dependent on the type of company, its management, its target market, and its overall mission.

*Stages of Development*

The Executive Summary should indicate three time-frames or stages of development. First, the "ramp up" period should show development prior to the first day of operations. For example, what growth or goals will be achieved within the 12 to 18 months prior to actual operations. Second, what development is expected to occur between the first day of operations and the end of year three? Third, what growth strategies are expected to occur from the end of year three to the end of year five? A timeline indicating growth beyond any more than five years will again, raise a red flag in the reader's mind. The environment in which the air-carrier will operate will most likely have changed significantly over a five-year period. Airlines operate with a relatively short-term vision versus an airport that operates with more of a long-term vision. There is a saying that airlines live for today while airports live for tomorrow.

*Financials*

The key financial assumptions and projections should be summarized in this section of the Executive Summary. A mention should be made regarding the income statement and associated expenses and revenue. The actual business plan will contain the Company's Operation Plan Summary for Years 1, 2, 3, 4, and 5 and the Executive Summary should highlight some important factors regarding this.

*Funds Sought and Utilization*

This section of the Executive Summary should indicate how much capital investment is required to launch operations or to expand existing operations and what the expected rate of return is. At the same time, the reader should be told what the required money will be used for and timelines for expenses might be discussed as well. For example, initial capital requirements might be used to complete air-carrier certification, to set up a reservation system, initial marketing and public relations campaign, securing of aircraft, recruitment and training of employees, establishment of contracts or agreements with vendors, prepaid expenses for fuel or insurance, or general corporate expenses.

**The "Do's" and "Dont's"**

When creating an Executive Summary, there are some important "Do's" and "Dont's" and it is important for the creator(s) to have a full understanding of the importance of such a document before it is distributed to potential investors or any interested parties.

*"Do's"*

- Be very cautious of who has access to the Executive Summary.
- Request that the original copy of the Executive Summary be returned to the owner upon completion of reading within a reasonable amount of time.

- When making copies of the Executive Summary available, be sure to make a limited number of copies to avoid mass distribution of sensitive information.
- If the reader requests to read the business plan upon completion of reading the Executive Summary, have the reader sign a Non-Disclosure Statement prior to making the business plan available (as discussed in *Chapter 4*).

*"Dont's"*

- Never write the Executive Summary before the business plan is completed.
- Do not provide all of the strategies that will be used to achieve success.
- Do not distribute the Executive Summary freely - only to credible parties.
- Do not allow readers of the Executive Summary to photocopy the document.
- Do not provide a copy of the Executive Summary and business plan together when presenting the Executive Summary for the first time.
- Do not make the Executive Summary available in an electronic format. Use a hard copy only.

## Summary

The Executive Summary, also known as a "teaser package", is the document created once the business plan has been fully developed. The Executive Summary is used to attract potential investors to the business and acts as a selling tool. The document summarizes the main elements of the airline's business plan without giving away all the secrets. If a potential investor is enticed by the contents of the Executive Summary, then the business plan can be made available for further interest. This chapter concentrated on the definition of the Executive Summary, described the main elements of the document, and briefly described the "Do's" and "Dont's".

## Key Terms

Executive Summary

Teaser Package

Return on Investment (ROI)

## Review Questions

1.  What is the product?

2.  What is the value proposition?

3.  What is the competitive advantage?

4.  What is the stage of development?

5.  What is the size of the target market?

6.  What is the revenue and profit model?

7.  Who are the team members and what is their contribution?

8.  Does this company have intellectual property or any assets?

9.  How much capital is needed and when?

## Web Sites

There are literally thousands of web sites available through the various search engines on the Internet describing how to write an Executive Summary. Many of these sites are private corporations attempting to sell a business plan product like a template or software program. There is no single source that the author of this book recommends for researching in the correct way to write such a document. The Executive Summary should include the main elements described in this chapter. By following the contents of this chapter, there should be no need to investigate any further. The author of this book reminds the reader to be cautious of business plan products because most products are "generic" in their design and are not appropriate for the airline business.

## Recommended Reading

Arkebauer, James. *The McGraw-Hill Guide to Writing a High-Impact Business Plan: A Proven Blueprint for First-Time Entrepreneurs*, McGraw-Hill, New York, 1994

Holloway, Stephen. *Straight and Level: Practical Airline Economics*. Burlington, VT, and Aldershot, UK: Ashgate Publishing, 1993.

Morrell, Peter.  *Airline Finance*.  Brookfield, VT: Ashgate Publishing, 1997.

Piotrowski, Maryann. *Effective Business Writing: Strategies, Suggestions, and Examples, a Guide for Those Who Write on the Job*, Harper Collins, New York, 1996

Tretheway, Michael and Oum, Tae.  *Airline Economics: Foundations for Strategy and Planning*.  Vancouver, Canada: University of British Columbia, 1992.

 # Chapter 6

## Business Introduction

**Chapter Checklist**

- Name, type of company, office location

- Business concept

- Passenger target market

- Destinations served

- Source of financial backing

- Current position of the company

## Introduction

Aside from the Executive Summary and the Non-Disclosure Statement, the Business Introduction is the first major element of the airline business plan. Although this section is somewhat short in length, it is rich in content providing important information for the reader. Under normal circumstances, this section of the business plan is written first and all details are initially presented. The Business Introduction outlines specific objectives and accomplishments, reveals barriers/problems/failures/risk, presents the "uniqueness" of the company, market attractiveness, and success factors all in a concise manner.

This chapter introduces the reader to the main elements of the business introduction including: name and type of the company, business concept, passenger target market, destinations served, source of financial backing, and current position of the company.

### Total Page Length: Approximately 1 Page

## Name, Type of Company, Office Location

In this section, the full name of the company should be presented along with the type of company and its office location, if applicable. If the office location is given, be sure to provide details about the facility such as lease expiry date, if lease is renewable, square footage, and its price. If the property is owned, describe the value and annual costs. Also, a mention should be made about any environmental concerns if they exist. The Business Introduction might start with the example below. It is important to remember that the business plan will be read by people familiar with aviation and others who are not. Be careful to avoid jargon or industry specific words in this section of the airline business plan in order to avoid confusion.

> Utopia Airways, LLC, is a Florida limited liability company organized to provide scheduled-charter passenger airline service to leisure destinations in Central Florida, the Caribbean, Latin America, and underserved destinations in the Northeast United States. Utopia Airways, LLC was incorporated in the State of Florida in May 200X. The offices for Utopia Airways, LLC are located at (full address).[1]

---

[1] Utopia Airways, LLC is a fictitious company designed specifically for the purposes of this book.

The "LLC" above refers to the legal form of the company. This is important to note in the Business Introduction so that the reader is aware of the legalities associated with this type of company. For example, in the United States, an LLC, or Limited Liability Company, gives the best of two business worlds by providing the informality and tax benefits of a partnership or sole proprietorship, and the personal protection from business debts of a corporation. There are many types of legal forms a company might consider but the LLC is most common when putting together the business plan. The legal form of the company can always be changed once the company is about to be established as an actual airline. In the United States, options range between a sole proprietorship, partnership, or corporation ("C" for standard or "S" for small businesses with limited shareholders). The format of the Business Introduction is suitable for most countries of the world but the legal form will vary country by country. The developer(s) of the business plan must research this aspect independently. There are many sources providing information on the different types of legal forms, including the Internet, but it is recommended that a lawyer specialized in the art of incorporation be consulted.

The Business Introduction should also provide a brief history of the company explaining the major events that led up to this point. List the key factors and the airline's niche within the industry. For start-up companies, this section of the business plan should list all the reasons why the business should be started.

## Business Concept

The reader should obtain a clear view of what type of business the plan proposes and should be able to answer the question, "What is this all about?" Very briefly, the Business Introduction must be able to answer the questions posed below.

- What type of product or service is being offered?
- What industry is the business plan aimed at?
- What are the needs and requirements of the customer?
- Why do these needs exist?
- Does the air carrier have the resources to develop, market, sell, and deliver the product or service?
- Are the resources required to develop, market, and sell readily available?

- How will the air carrier communicate the product and its advantages to potential customers?
- What are the significant differences between you and the competition? (For example, technical, positioning, segmentation, etc.)
- How will advantages be sustained over time, given the competitive nature of the industry?
- What has been the air carrier's market experience to date? How much contact has already been made with prospective customers and what type of reaction has been experienced?

**Passenger Target Market**

This section of the Business Introduction describes the product and service provided by the airline. There is some confusion in the airline industry as to whether airlines offer a product or a service. For the purposes of this book, airlines are both product and service oriented. Passenger airlines, regardless of size, offer a product known as a seat. The airline simply rents a seat to the passenger for the purpose of moving the passenger from one destination to another destination. Anything beyond the rental of a seat is considered service including the contact between the airline and the passenger, meals and beverages regardless if complimentary or paid for, entertainment, executive lounges, etc. In terms of air cargo, many passenger airlines offer this as an extra service because it is not the core business. For air cargo carriers, the product offered is the space made available for rent by the customer. Service is considered to be the interaction between the air carrier and the customer and anything else beyond the rental of space on the aircraft.

When designing an airline business plan, one should know that there is not one market for passengers but many. An airline attempting to establish itself in all the markets will most likely fail. Historically, many major air carriers have tried to market themselves in too many different types of markets without focusing on a specific niche. Today, airlines are facing restructuring barriers and are realizing that it is more beneficial to target specific types of markets in order to keep costs down. New airlines entering the industry are attempting to establish a strong presence in niche markets where there is passenger demand and little to no competition. Current trends indicate that underserved secondary markets hold the greatest potential compared to highly competitive routes connecting major city centers with major city centers. Examples of possible passenger target markets include: leisure, business, budget, students, military, secondary city to secondary city, secondary city to major city, major city to secondary city, major city to major city. The business plan should clearly define what type of passenger(s) the airline is aimed at. If the passenger target market is not known, then there is no sense in writing the business plan.

## Destinations Served

At this point, the airline has confirmed the passenger target market therefore confirming the route network. Specific locations do not have to be defined in this section but the general areas or regions should be noted as shown in the example at the start of this chapter. The actual destinations are discussed later in the business plan under the section entitled, *Proposed Route Structure and Schedule.*

## Source of Financial Backing

If a business plan is being designed for the first time, there is a small chance that no to little money has been raised to help launch operations. The Business Introduction should familiarize with the current position and what plans have been made to raise the required capital. The statement below is sufficient for this portion of the business plan as more detail is given concerning raising of capital later in the business plan.

> As of May 200X, Utopia Airways, LLC is in the process of raising capital towards providing the necessary infrastructure, operating contracts, and government certificates to allow for the commencement of scheduled passenger service. It is the opinion of Utopia Airways management, current investors, and interested parties that with the required level of initial capital, the Utopia Airways business strategy will provide efficient airline operations, offering excellent service to its customers at competitive fares, allowing Utopia Airways to be a long-term success within the airline industry.

## Current Position of the Company

This section of the Business Introduction should outline the current position of the company in terms of where it is at "today" and where it plans to be "tomorrow". A time frame should be outlined so the reader can see what anticipated progress is expected. Current ownership of the company might be mentioned at this point, including a list of all shareholders and percent of ownership. A brief description of each owner should be presented describing his/her position and area of expertise. Shareholders who are not involved in the day-to-day operation of the airline should be noted as well as they may be required to obtain independent legal advice prior to signing a personal guarantee for a bank loan.

In general, the people involved in the design of the business plan will not be the people to hold the main executive positions once the company begins to raise capital. Many people dream of having "their" own airline but lack the funds to make the dream come true. Unfortunately, as the airline business plan is marketed to potential investors, the people with money become more powerful than those with the initial idea and have a greater say in terms of who will do what. In order to protect one's self from being "thrown" out of the picture, it is possible to secure a compensation package in exchange for information. Some investors may consider offering a buy-out deal or compensation in the form of shares or profit sharing. Perhaps the most frustrating part of creating a solid business plan is not being able to maintain control once things start to become a reality. Of course, this is not applicable to all cases.

## Summary

The Business Introduction is the first main element of the airline business plan. This section is very important despite being short in length. The amount of content provided gives the reader a good idea of exactly what the business concept proposed is all about. The company's objectives and accomplishments are summarized as are potential barriers and risks. The Business Introduction explains to the reader why the company is unique compared to the competition and discusses how the airline will achieve success by setting itself apart from others in terms of market attractiveness and business strategies. A number of questions were posed in a logical format in this chapter and each must be considered and answered fully if the business plan is to be successful. The Business Introduction should be written first before any other section of the business plan but should be expected to be revised once all the other elements of the airline business plan have been researched and completed in full.

## Key Terms

Limited Liability Company (LLC)

Proprietorship

Partnership

Corporation

Product

Service

## Review Questions

1. What is the primary purpose of the Business Introduction and what are the main factors that must be identified?

2. Does my Business Introduction explain the name, the location, and the type of company being proposed?

3. Have all the questions identified under the business concept been asked and answered?

4. Is the passenger target market well defined and are the right airport sites being utilized in the decided destinations?

5. What are the sources of financial backing and what is the current position of the company?

## Web Sites

http://www.sba.gov/starting_business/planning/basic.html

http://www.businesstown.com/planning/creating.asp

http://www.bplans.com/ho/index.cfm

http://www.planware.org/freeware.htm

http://www.users.cloud9.net/~kvivian/html/business_plans_.html

## Recommended Reading

Abeyratne, Ruwantissa. *Aviation Trends in the New Millennium.* Burlington, VT: Ashgate Publishing, 2004.

Arkebauer, James. *The McGraw-Hill Guide to Writing a High-Impact Business Plan: A Proven Blueprint for First-Time Entrepreneurs*, McGraw-Hill, New York, 1994

Graham, Brian. *Geography and Air Transport.* Chichester, UK: John Wiley & Sons, 1995.

Holloway, Stephen. *Straight and Level: Practical Airline Economics.* Burlington, VT, and Aldershot, UK: Ashgate Publishing, 1993.

Morrell, Peter. *Airline Finance.* Brookfield, VT: Ashgate Publishing, 1997.

Piotrowski, Maryann. *Effective Business Writing: Strategies, Suggestions, and Examples, a Guide for Those Who Write on the Job*, Harper Collins, New York, 1996

# Chapter 7

## Mission

**Chapter Checklist**

- Mission Statement

- Vision Statement

## Introduction

There are many factors leading to the success and failure of airlines of all sizes around the world. One such factor is the Mission Statement. The Mission Statement is a must for every airline, regardless of its size because it will aid in achieving its ultimate success. Unfortunately, one common mistake is frequently made in the airline business when it comes to knowing what the mission of the company is. The goal of the Mission Statement is to define the main purpose or the aim of the airline. Once this has been done, the Mission Statement must be backed up by a Vision Statement and this is one variable that is often left out of the airline business plan. The Vision Statement outlines the airline's desired destination within a specific time-frame. The need for flexibility within the airline business plan was discussed in an earlier chapter. Flexibility must be incorporated into the Mission and Vision Statements to accommodate changes in the environment in which the airline competes. This chapter discusses the importance of the Mission Statement, Vision Statement, and introduces the reader to the key elements of each.

**Total Page Length: 1 Page**

## Mission Statement

In the airline industry, failure is not an option. Success is often dependent on the establishment of a Mission Statement designed to clearly define the main purpose or the aim of the organization. The Mission Statement must be long lasting and flexible to accommodate internal and external challenges created by changes within the environment in which the airline operates. In other words, to take the airline to the next level, stability must be created through the Mission Statement. Once the mission has been determined, it is important for the wording of the Mission Statement to remain the same but changes in terms of how the airline operates can be accommodated without requiring the Mission Statement to be altered. One might ask how stability and flexibility can be combined. This will be discussed in the section where the Vision Statement is introduced.

The wording of the Mission Statement depends entirely on the airline's structure. In order to define the company, the core competency of the airline must be examined by asking the following questions:

1. Who are you?
2. What do you do?
3. For whom and what is your uniqueness?

Some organizations find it beneficial to have more than one Mission Statement each aimed at a specific group. Airlines are often unstable when it comes to their financial and labor issues and any tool that will help provide a more stable environment is welcomed. When designing the Mission Statement, an airline might find it useful to have separate statements for each of the following groups: passengers, employees, and stakeholders. Providing direction for each group will help the airline meet its overall goal.

Once the Mission Statement is completed, the airline will find it to be an important marketing tool that will help the company communicate with passengers and the various organizations involved in airline operations by telling everyone concerned the reasons for the airline's existence. It is a tool for leadership because it gives the airline a clear position to use when working with allies and promoting the airline with the public. The Mission Statement is a motivational tool because it encourages employees and allies to work toward a common vision.

The Mission Statement should be brief ranging from a few words to a few sentences. Brevity will enhance understandability but be cautious not to be too general as it is very important that the intended purpose of the organization be clearly defined.

The Mission Statement should be easily remembered using simple words that people can actually use.

The Mission Statement should be audible meaning that the use of words and phrases should sound good. It is often said that a good Mission Statement will use a conversational tone.

The Mission Statement should be "becoming" in the sense that it states what the airline is as well as how it will reach its goals.

The Mission Statement should be unique and be able to demonstrate how the airline differs from its competition.

A well designed mission will make a positive difference and accomplish the following:

- Increase financial performance of the airline (revenue, profit, market share).
- Inspire organizational members, harness their energy and passion and increase their commitment to achieving goals and objectives.
- Increase organizational focus through the sense of purpose and direction that are shared throughout the airline.
- Distinguish the airline from its competition.
- Challenge the airline to achieve its objectives.
- Enhance decision-making by harmonizing values and vision.
- Improve the planning process.

**Examples of Mission Statements**

The following Mission Statement is an example designed solely for the purpose of this book. The remaining examples are working mission statements produced by existing airlines.

*X Airways* - Passengers

The mission of *X Airways* is to provide underserved markets with affordable, high quality, international, scheduled-charter air service to leisure destinations. *X Airways* will stimulate community economic growth by serving underutilized secondary airports while providing direct access to high demand markets, on routes not currently served by any other major air carrier. *X Airways* plans to lead the industry in the use of the latest modern technologies, in combination with old fashioned, superior and personalized customer service, in order to provide a reliable and pleasurable travel experience.

**X Airways** - Employees

> *X Airways* aims to create a challenging and exciting environment in which creative people can work as a team. We employ only the best and the brightest, and look to recruit only those with education and experience that makes them a cut above the rest.

**X Airways** - Shareholders

> It is our mission to ensure that our shareholders see a fair return on their investments in *X Airways*. By empowering our employees with the latest training, *X Airways* continues to raise the level of excellence and service we provide to our passengers. Our talented professionals are also vested shareholders in the company they help build.

### *American Airlines*™

> American Airlines™ seeks to be the global leader in incentive travel. We will accomplish this through aggressive marketing on a global scale, developing and maintaining partnerships with our corporate clientele, and providing superior customer service and support of our products.

### *easyJet*™

> To provide our customers with safe, good value, point to point air services. To effect and to offer a consistent and reliable product and fares appealing to leisure and business markets on a range of European routes. To achieve this we will develop our people and establish lasting relationships with our suppliers.

### *KLM Royal Dutch Airlines*™

> KLM wants to excel in the quality of its connections, by linking as many cities as possible. KLM's goal is to participate in one of the leading global airline systems as an independent and financially strong European partner. KLM aims to sustain passenger preference and provide a stimulating and stable working environment for its employees. Targets also include a structural increase in shareholder value and a mutually profitable relationship with its partners.

*SkyWest*™

> Our mission is to provide airline service that exceeds our customers' expectations. Each of us is dedicated to excellence in the air transportation system. We are dedicated to: Safety first; On-time every time; Fairness and Consistency in everything we do; Working together as a team; Personal and corporate integrity; Maintaining profitability.

*Southwest Airlines*™ - Passengers

> The mission of Southwest Airlines is dedication to the highest quality of Customer Service delivered with a sense of warmth, individual pride, and Company Spirit.

*Southwest Airlines*™ – Employees

> We are committed to provide our Employees a stable work environment with equal opportunity for learning and personal growth. Creativity and innovation are encouraged for improving the effectiveness of Southwest Airlines. Above all, Employees will be provided the same concern, respect, and caring attitude within the organization that they are expected to share externally with every Southwest Customer.

## Vision Statement

> The Mission Statement cannot be completely effective unless followed by a Vision Statement. The main purpose of the Vision Statement is to define the airline's future success within a specific time-frame. In other words, today's vision equals tomorrow's reality. This important aspect of the airline business plan should only be created once the Mission Statement is known.

> As discussed previously, the Mission Statement helps provide for stability within the airline while the Vision Statement provides flexibility. This section of the airline business plan is a detailed and powerful piece of communication creating a sense of desire and it builds commitment to reaching the vision.

> The Vision Statement should not be copied from another company, otherwise the airline will lose the uniqueness it is attempting to acquire. The Vision Statement should paint a complete picture of what the future will look like while instilling rigorous discipline in the process of developing the vision. It should be realistically achievable and effectively communicated to all the parties concerned. Just like the Mission Statement, the Vision Statement is

aimed at passengers, employees, and stakeholders. When recruiting for employment positions or seeking investors, the airline should seek inspired people that share a common dream.

The Vision Statement should start correctly the process of Successful Change Management while embarking in the development of strategies that will lead to success. A successful Vision Statement will empower employees and provide a corporate culture that makes it easier for everyone to make the right decisions the first time around. Once the implementation process is initiated, it is important to keep track of the airline's vision and make unproductive behavior patterns that are not aligned with the vision stand out. Keep track of challenges to the airline's vision and fully understand challenges. This will aid stakeholders to communicate to management challenges to the vision. The Vision Statement is flexible compared to the Mission Statement, because it allows modifications to be made as the internal and external environment changes. In other words, a successful Vision Statement manages change efficiently.

## Summary

There are a number of factors leading to the success and failure of airlines around the world. One problematic factor for many airlines is that although they have a mission, they often lack a vision within the airline business plan. This chapter introduced the reader to the Mission Statement and defined it as the portion of the business plan that clearly states the airline's main purpose or aim. The Mission Statement helps provide stability for the airline while the Vision Statement provides flexibility permitting modifications where necessary given changes in the environment in which the airline operates. The Vision Statement is detailed providing a clear picture of what the airline's success will look like within a specific time-frame.

## Key Terms

Mission Statement

Vision Statement

## Review Questions

1. What is the purpose of a Mission Statement and is it long-lasting and flexible?

2. Does the Mission Statement clearly define who we are as a company? In other words, does the Mission Statement answer all the questions necessary to determine our core competency?

3. Is one Mission Statement suitable for our operation or is a separate statement required for each participant within the airline's structure? In other words, should a separate Mission Statement be created for the passengers, the employees, and the stakeholders?

4. What is a Vision Statement and how much flexibility does it permit given changes within the internal and external environment in which the airline operates?

5. Is the Vision Statement realistically achievable and effectively communicated to all participants within the airline's structure?

## Web Sites

http://www.corporatemissionsinc.com/index.htm

http://www.business-plan-success.com

http://www.bizplanit.com

http://www.businessvision.co.uk

## Recommended Reading

Abrahams, Jeffrey. *The Mission Statement Book: 301 Corporate Mission Statements From America's Top Companies.* Berkeley, CA: Ten Speed Press, 1995.

Freiburg, Kevin and Jackie Freiburg. *Nuts! Southwest Airlines' Crazy Recipe for Business and Personal Success.* New York, NY: Broadway, 1996.

Haschak, Paul G. *Corporate Statements: The Official Missions, Goals, Principles and Philosophies of Over 900 Companies*. Jefferson, NC: McFarland, 1998.

Jones, Patricia. Say *It & Live It: 50 Corporate Mission Statements That Hit the Mark*. New York, NY: Doubleday, 1995.

# Chapter 8

## Strategy

**Chapter Checklist**

- Kind of airline

- Strategic Action Plan

- Strategy implementation

- Growth and expansion strategy

- Failures in planning

## Introduction

Within the airline business plan, a strategic plan is also necessary. The strategic plan acts as a "game plan" describing the overall approach or plan of action for achieving growth and profits. The strategies outlined are based on the general business concept. Historically, many airlines implemented effective strategies but they were designed for the short-term and not the long-term. Daily changes occur within the airline business and it is important that strategies cater to the constantly changing environment.

To be successful in today's aviation environment, strategies must be designed to be flexible allowing the carrier to adapt to the changing environment. In the twenty-first century, this is proving to be difficult but not impossible. This chapter introduces the reader to specific key steps to success and discusses topics like the strategic action plan, strategy implementation, growth and expansion strategy, and failures in planning.

**Total Page Length: Approximately 2 to 3 Pages**

## What Kind of Airline?

An airline must know what type of business is to be operated or is currently being operated before the concept of strategies can be discussed. For example, is the airline scheduled or non-scheduled? Is the airline to operate as a shuttle (corporate or commercial)? Will the airline offer ad hoc charters? Is the airline domestic and/or international? Will the airline offer short haul, medium haul, long haul routes or a mixture? What is the passenger target market? Will the airline offer no-frills low-cost, low-cost, medium grade, or premium grade services? Will the airline subcontract work or perform all of its operations in-house? All of these questions must be answered in detail before the strategic action plan can be developed.

## The Strategic Action Plan

The strategic action plan's primary objective is to guide management and make sure they do not stray too far off from the main objectives of the company. The main topics include: airline's reason for existence, main objectives, position in the market, how the airline will compete, how the airline will satisfy customers needs, and how the airline will achieve good business performance. The strategic action plan should be tailored to the type of business the airline plans to compete in. An airline strategic action plan might include the following headings in order: meet the desires of the customer base, selection of optimal aircraft fleet, base of operations, outstation markets and description, and route strategy. The media budget and plan should be included within the strategic action plan as well. Incorporated into the suggested headings above, six key evaluation questions should be answered within the strategic action plan:

1. Are aspirations consistent with the external environment?
2. Is the opportunity appropriate given the available resources?
3. Is there an acceptable degree of risk?
4. Is there an appropriate time-frame?
5. Is the opportunity workable?
6. Does it motivate?

## Strategy Implementation

A good model to follow in terms of successful implementation of strategy, is the former U.S.-based Reno Air. When Frank Lorenzo was designing the business plan for this airline, he acknowledged six key factors known as the strategic steps to success.

1. Figure out the best location
2. Establish a top management team
3. What, if any, services to offer?
4. Stay flexible and be opportunistic
5. Design an original marketing plan
6. Admit mistakes

Once management has designed the appropriate strategies for the strategic action plan, there are eleven main elements of strategy implementation that should be followed. The six key factors listed above combined with the following eleven elements below, should lead to successful operation if implemented properly.

1. Once the strategy is finalized, the key tasks to be performed and kinds of decisions required must be identified.

2. Once the size of operations exceeds the capacity of one person, responsibility for accomplishing key tasks and decision-making must be assigned to individuals or groups. For new airlines starting out, it is important to start with a small management team with multiple responsibilities. Historically, airlines, specifically major carriers, have been guilty of having too many managers resulting in duplication of duties.

3.  Formal provisions for the coordination of activities thus separated must be made in various ways (hierarchy, committees, task forces).

4.  Information systems adequate for coordinating divided functions must be designed and installed. Today, airlines are becoming more virtual and modern airlines are finding it beneficial to invest into new Information Technology (IT) rather than into people. However, this does not mean that technology should replace people entirely.

5.  The tasks to be performed should be arranged in a time sequence comprising a program of action or a schedule of targets. Airlines of all sizes should have written procedures in place to improve efficiency. This is especially true as the airline grows and new departments or divisions are added.

6.  Actual performance should be compared to budgeting performance and to standards in order to test achievement, budgeting processes, and the adequacy of the standards themselves.

7.  Individuals and groups of individuals must be recruited and assigned to essential tasks in accordance with the specialized or supervisory skills which they possess or can develop. Historically, many airlines have been guilty of not hiring the "right" people for the "right" job.

8.  Individual performance should be subjected to influences, constituting a pattern or incentives, which will help make it effective in accomplishing organizational goals. Airlines should develop methods to motivate employees by improving corporate culture.

9.  Since individual motives are complex and multiple, incentives for achievement should range from those which are universally appealing to specialized forms or recognition designed to fit individual needs and unusual accomplishments.

10. Provision for the continuing development of requisite technical and managerial skills is a high-priority requirement.

11. Dynamic personal leadership is necessary for continued growth (improves achievement in any organization).

**Growth and Expansion Strategy**

Although many airlines "live for today", strategies should be implemented within the strategic action plan that allow the airline to "live for tomorrow". Investors will be reluctant to invest money into an airline if there is no vision for the future in terms of growth. A suitable time frame is three to five years with five years being preferred. This section of the strategic action plan should provide the reader with a clear view on how management will sustain the business.

Once in the market, management should be able to describe how the opportunity will be grown and at what point the airline will hit profitable status with acceptable market share. The following questions must also be answered. Are there any spin-off or related opportunities that can be introduced later to further grow the opportunity? If so, how will this be accomplished? Which customers and segments will the airline depend on for its sustained growth? How will the airline expand geographically – if, where, when? What advertising, marketing, and promotional approaches will management use to sustain market share over the long haul after initial introduction? Forms of advertising, marketing, and promotional strategies include:

- Advertising (print, radio, television)
- Direct Mail Communications
- Telephone Marketing
- Promotional Events (trade shows, seminars, workshops)
- Publicity and Public Relations (news releases, articles, speeches, industry associations)
- Sales Promotions

Today, the Internet is the preferred method for ticket sales. This is referred to as direct marketing because there is no middle person involved adding to operating costs. More and more airlines are encouraging passengers to utilize the Internet by accessing the airline's home page to obtain information and book tickets. The Internet permits 24-hour access without the need for staff and expensive infrastructure. Some airlines are now 100% virtual in the sense that a passenger can only fly with a specific airline if the ticket is purchased over the Internet. Advertising the Internet address for the airline on the side of the aircraft is becoming increasingly popular.

## Failures in Planning

When it comes to the failures of strategic planning, the airline industry is no different than any other type of business. Management must recognize the following pitfalls and implement appropriate solutions to avoid such mistakes.

- Management assumes planning can be delegated to a planner
- Wrong strategy selected to achieve organization objectives (not flexible)
- Wrong organizational design used to implement the strategy
- Preoccupation with current day-to-day activities
- Failure to develop suitable company goals
- Failure to involve line personnel
- Lack of motivation
- Failure to use plans as standards for measuring performance
- Too formal; lack of flexibility; looseness; simplicity
- Ineptness or negligence on the part of management
- Failure to review plans adopted/finalized
- Opting for intuitive decisions which conflict with the formal plans
- Lack of communication within the organization

## Summary

This chapter discussed the strategic action plan, strategy implementation, growth and expansion strategy, and failures in planning. The strategic action plan is a part of the business plan and is used to guide management with daily decisions impacting both short-term and long-term operations. It was acknowledged that many airlines have made the mistake of not planning for the future while concentrating too much on the short-term. As the aviation environment rapidly changes due to external forces, airlines find it difficult to succeed because their business and strategic plans are not flexible. The strategic action plan should answer six key questions based on the type of airline being designed. Once the strategies have been developed, they should be considered alongside six strategic steps to success followed by eleven key elements of implementation.

## Key Terms

Strategy

Strategic Action Plan

Flexibility

## Review Questions

1. What is the main purpose of the strategic action plan?

2. What type of airline is the strategic action plan to be designed upon?

3. What are the six key evaluation questions asked in the strategic action plan? Does my strategic action plan do a thorough job of answering these questions?

4. What are the six key factors known as the strategic steps to success? What are the eleven main elements of strategy implementation? Does my strategic action plan incorporate the key steps to success and the elements of strategy implementation?

5. There are a number of common failures in strategic planning. What are the main failures and does my strategic action plan do a thorough job in avoiding these failures?

## Web Sites

http://www.corporatemissionsinc.com/index.htm

http://www.business-plan-success.com

http:///www.bizplanit.com

http:///www.businessvision.co.uk

## Recommended Reading

Dempsey, Paul Stephen and Gesell, Laurence. *Airline Management: Strategies for the 21$^{st}$ Century.* Chandler, AZ: Coast Aire Publications, 1997.

Piotrowski, Maryann. *Effective Business Writing: Strategies, Suggestions, and Examples, a Guide for Those Who Write on the Job*, Harper Collins, New York, 1996

Radnoti, George. *Profit Strategies for Air Transportation*. New York, NY: McGraw Hill, 2002.

Tretheway, Michael and Oum, Tae. *Airline Economics: Foundations for Strategy and Planning*. Vancouver, Canada: University of British Columbia, 1992.

Wells, Alexander T., and Wensveen, John G. *Air Transportation: A Management Perspective, 5th ed.*. Belmont, CA: Thomson – Brooks/Cole™, 2004.

# Chapter 9
## Market Opportunity

**Chapter Checklist**

- Type of service

- Market analysis and opportunity

- Airports to serve

## Introduction

Historically, airlines have not done a good job when it comes to market research concerning route networks. The type or types of passengers the airline serves determines specific routes, therefore, determining specific airports the airline will operate at. Many airlines have failed because of the poor quality of their research. In the United States, prior to the U.S. Deregulation Act of 1978, airlines did not have a need to do research because there was almost no competition. In other words, airlines had a monopoly on certain routes and passengers were forced to fly certain airlines regardless of price or desire. The same applied to many of the European Union (E.U.) countries until the mid-1990s when the Third Package was implemented. The Third Package was the final step of creating a liberalized environment in Western Europe where carriers can fly to any destination, at any price, and compete with other carriers as long as they are operating in a safe environment.

In the twenty-first century, airlines around the world are finding that extensive research concerning passengers and destinations is required, due to an increasingly competitive environment. Many airlines now spend great portions of their annual budget on market research because airlines have realized for the first time in their existence that passenger loyalty no longer exists. Passengers will fly with the carrier that provides the best price and gets them to their end destination on time. In today's aviation environment, passengers are price sensitive whereas before, passengers were more time sensitive.

Even though extensive market research is necessary, there is no guarantee that the airline will be successful. Forecasting techniques are simply forecasts and the only real way to "test" a market is to operate an actual aircraft on a route. If successful, the airline has virtually no worries. However, if this test is not successful, the airline must have a contingency plan in place to determine how the aircraft will be utilized without it spending time on the ground. The author of this book believes three trends are occurring within the global airline industry. It is important for the developer of the airline business plan to be able to identify future trends and select a specific tier in terms of what category of airline the business plan is to be designed around. The three tiers include: regional/feeder carrier, new-entrant/low-cost/no-frills carrier, and the "megacarrier". This chapter introduces the reader to discovering potential market opportunities.

**Total Page Length: Approximately 4 Pages**
**(Depending on number of destinations served and amount of information presented)**

## Type of Service

As noted earlier in this book, many airlines are guilty of not doing thorough research when it comes to researching route structures and passenger types. A good business plan incorporates extra effort when it comes to background research. However, finding a point to start can often be a difficult task. One question should be answered before any research is undertaken: "What route(s) do you have in mind?" To investigate this question, the developer should research underserved routes, saturated routes, become familiar with passenger demand (past, today, tomorrow), and also become familiar with destination legalities. The latter factor is most important when dealing with destinations outside the airline's home country.

Many airlines have evolved based on a "hunch". Some have succeeded while other have failed. Starting an airline is a risky business as it is and anything the developer can do to minimize risk is most beneficial to achieving success. Having a general idea of what kind of airline one wants to start is good but background research is necessary to make sure the idea is sound. In most cases, the developer will find new opportunities or barriers that weren't known in the preliminary stages creating a new concept.

In the airline industry, there are many different types of services that can be offered and the business plan should be focused on a particular area or niche market. In most cases, the more specialized the airline is, the better the chance of success. Examples of different areas to research are presented below. It should be noted that this is merely a list of examples to get started and not a complete listing of all the areas requiring research. Once a thorough investigation has been done, the developer will have a greater understanding of what type of airline will be designed in the airline business plan.

### Scheduled or Non-Scheduled Service

For the most part, an airline will offer either a scheduled service or a non-scheduled service. A scheduled airline will fly to different destinations using a published time schedule. For example, *X Airways* offers service from *Airport A* to *Airport C* on Mondays, Wednesdays, and Fridays departing at 0700. Depending on the country of registration, the airline will operate under a particular flight certificate authorizing scheduled service. This certificate is issued by the government (civil aviation authority) of that country. A non-scheduled airline will offer services to different destinations but will not fly according to a published time schedule. For example, *X Airways* offers service from *Airport A* to *Airport D* but the days and times might not be specific.

Again, depending on the country of registration, the airline will be issued a specific flight certificate authorizing non-scheduled service.

*Luxury, Mid-Range, Low-Cost and No-Frills, Shuttle, and Charter*

When building the airline business plan, the developer must know what type of service to operate in terms of the amenities it will offer. Generally, a luxury-oriented airline stands a good chance of failing from the start due to high overhead costs. A good example of this type of carrier was the U.S.-based MGM Grand Air that provided luxury service between New York and Los Angeles. One of the predominant reasons of its failure was the high overhead costs associated with offering First Class seating with extra legroom, china dishes for meal service, exotic food and drink, and aircraft that were expensive to operate and maintain.

A mid-range airline will cater to passengers wanting a reasonable airfare with some in-flight amenities including food, drink, and entertainment. Generally, mid-range airlines have a reasonable chance to survive as long as the cost structure is well maintained. For the most part, major airlines are categorized as mid-range.

Low-cost carriers cater to passengers wanting cheap airfares with little demand for in-flight services. However, it is important to distinguish between a low-cost and a no-frills carrier. In terms of cost structure, a low-cost airline offers a reasonable airfare resulting from low-cost management strategies. A no-frills airline also offers reasonable or cheap airfare resulting from what might be considered extreme low-cost management strategies. Basically, a no-frills airline offers a seat from point A to point B with no in-flight service. In the United States, Southwest Airlines is considered the leading low-cost no-frills air carrier.

A shuttle airline caters mainly to business travelers seeking movement between two major city centers. The shuttle concept is similar to a conventional bus service offering a reasonable airfare with no reservation. High frequency and easily remembered times are typical attributes of a shuttle.

A charter airline offers services to destinations based on demand without using a published time schedule. In other words, the aircraft might be rented one time or multiple times to transport people or goods to specific destinations. This type of service is referred to as an ad-hoc charter. The more common type of charter caters to passengers seeking leisure-oriented destinations. Most airlines in the charter market operate by a non-published time schedule to specific destinations on a seasonal basis. In the northern hemisphere, many charter carriers operate north and south during the winter and east and west during the summer.

*First, Business, Economy*

What type of seating should the airline offer? Historically, major airlines offered three types of seating configurations: first, business, and economy. Today, as it becomes more difficult to operate a successful airline from a financial perspective, many airlines are doing away with the three classes and moving toward two classes. First class is being removed and replaced with increased business class seating and increased economy class seating. The marketing geniuses at many major airlines have renamed the expanded business class first class for psychological reasons. Increasing deck capacity means more revenue is generated for the airline. In short, "putting more butts in the seats", is the new trend. If an airline offers a first class seat, extra space is occupied by the seat because the first class passenger demands extra leg room meaning increased seat width and pitch. Alongside extra room, this particular passenger type demands a costly in-flight service consisting of food, drink, and personal entertainment.

Business class seating is important to airlines wanting to attract business travelers willing to pay a high air fare. Historically, business travelers have been time sensitive and not price sensitive meaning that the major airlines could offer a last minute seat and expect to generate a high yield. However, since the events of 9/11, a new trend has occurred.

Economy class seating is more important than ever before. Since 9/11, many corporations have cut down on operating costs by reducing travel budgets for employees. Business travelers that once traveled in business class are being forced to travel in economy meaning that many of the major airlines are no longer receiving the same high yields they previously did. For the first time in aviation history, major airlines are realizing that their "bread and butter" are economy class passengers. Low-cost airlines typically offer a single economy class and generate revenue based on volume rather than by seating class. An aircraft can accommodate more seats with a single seating configuration meaning that airlines operating with maximum deck capacity have lower operating costs passing the difference on to the passenger resulting in a reasonable ticket price.

*Food and Bar*

Deciding to offer food and drink on board an aircraft is becoming a more important decision than ever. Every time a meal or drink is served, it costs the airline money. More and more airlines are finding it beneficial to charge the passenger if he or she wishes to eat or drink. This is most noticeable with airlines of all types and sizes operating short and medium haul routes. Airlines flying long haul routes generally offer complimentary food and drink but in many cases, the volume and quality have decreased in order to cut costs. The amount of profit that can be generated from such sales is phenomenal.

*Entertainment*

Although many aircraft are equipped with various types of inflight entertainment, offering such entertainment can be a costly decision. The technology associated with offering movies, radio, television, and telephone is very costly and someone has to pay. Historically, such entertainment was complimentary. Just like food and drink, entertainment is now offered by many airlines for an extra charge. Many airlines have realized the profit to be made by selling headsets to the passenger.

*Cargo and Freight*

If a passenger airline plans to offer cargo or freight service, there are some important factors to consider. Firstly, due to height and weight restrictions, it is important that the transport of such goods does not interfere with the primary revenue generator – the passenger. Also, the type of aircraft operated will impact the amount of cargo and freight that can be hauled. A wide body aircraft is necessary to offer a pallet and container system. Currently, there is only one narrow body aircraft equipped to handle a pallet and container – the Airbus A320. Prior to any operations, the airline business plan should identify how much involvement with cargo and freight the airline plans for the future. The answer to this could impact the type of aircraft flown having a significant impact on costs. Current trends indicate that the transport of cargo and freight is beneficial if done on a supplemental basis. A passenger airline cannot compete with the mainstream door to door operators like Federal Express, UPS, TNT, and DHL.

*Duty-Free*

The offering of duty-free goods only applies to airlines flying on an international basis. For the most part, offering such a service is a positive move because it makes two parties happy. Firstly, the passenger appreciates the opportunity to purchase duty-free goods on board the aircraft. Secondly, the airline benefits by earning a profit on each sale.

*Baggage Restrictions*

As mentioned previously, many airlines are increasing deck capacity with increased economy and/or business class seating. Due to increased passenger weight, airlines are finding that they have to limit the amount of baggage a passenger can check-in. Passengers checking-in baggage beyond the airline's restriction are often charged an excess baggage fee. Many airlines have realized how much revenue can be earned as a result of such a fee.

*Interline Agreements*

Because the major airlines operate on a hub-and-spoke system, they have the ability to offer interline agreements with other airlines. Although this is positive from a marketing perspective, one of the potential downfalls is that interlining can cause contractual nightmares and baggage transfer headaches. It is also important to note that when interlining passengers, often the passenger is not aware of what airline they are flying on. As a result, bad service offered by one of the interlining airlines can be associated with your airline. On the other hand, a positive experience may also be associated with your airline. New-entrant airlines are finding it beneficial not to interline with competing carriers in order to keep operations simple. Also, because many new-entrant and low-cost airlines do not utilize hub airports, there is no reason to interline passengers because there is no need for connectivity.

*Other Amenities*

To increase an airline's chance of success in an increasingly competitive industry, the carrier should be able to offer amenities that competing carriers do not offer. Adding a sense of uniqueness will become more important as the twenty-first century progresses.

**Market Analysis and Opportunity**

The market opportunity of the airline business plan should give the reader a clear view of the market opportunity for the product and/or service. This section answers the question, "What is the nature, size, and life span of this opportunity?" It is important to note that the main product of an airline is a seat. Each seat should be treated as a perishable product similar to a fruit or vegetable. There is a specific period of time to sell the seat and once the aircraft door is closed, the seat has no further value. This section of the airline business plan should accomplish the following:

- Estimate the market potential for the product and/or service by (a) specific geography – how measured? and (b) estimated growth rate – how measured?

- What is the probable range of market share attainable, given the competition and passenger demand?

- Describe the range of profitable revenue potential based on reasonable market share.

- Describe how the airline's advantages and resources allow for significant profitability.

- How dynamic is the market? What is the growth rate? Is it flat? Is it declining?

- What is the probable life cycle of the product and/or service?

- How will the airline enter the market on a time scale that makes sense relative to the life cycle? Describe how the airline will meet the "window of opportunity".

- Describe related possible future product and/or service offerings that can expand the market potential and/or extend the life cycle.

- Are there unusual opportunities for additional differentiation through technology, positioning, or segmentation?

- How easy is it to enter the market?

- Describe the reaction or sales response from customer contact, interviews, and any field tests or trials conducted.

This section of the airline business plan might also include historical data, trends, geographical information, tables, and graphs related to each potential destination.


### Airports to Serve

Once the initial route structure has been settled, it is important to identify the airports the airline plans to serve. A home base and administrative location should be identified. When researching specific airports, it is important to make sure each location has the necessary facilities required for the type of airline operation. Also, it is important to make sure the airport can accommodate the time schedule the airline plans to operate. Night and noise curfews become important considerations as do runway length, taxiway design, parking infrastructure, customs and immigration facilities, and ground transportation.

In some regions of the world, airports operate on a slot system meaning that the airline must purchase an allocated piece of time to land and depart. In the United States, there are only four slot allocated airports (JFK, La Guardia, Chicago O'Hare, and Washington National). However, in Europe, virtually all major airports are slot allocated.

When selecting what airports to operate at, it is important to have an understanding of the different types of airports available. Becoming familiar with the advantages and disadvantages of a hub-and-spoke system is just as important as becoming familiar with secondary airports and the point-to-point system. Knowledge of both systems is necessary to fully understand network and cost strategies.

**Summary**

Discovering market opportunities is a challenging task requiring a lot of background research. Many airlines are guilty of not doing thorough research when it comes to network strategy. However, this old trend is changing as the competitive nature of the airline business increases. Additional research is also being spurred by pressure on airlines to reduce costs wherever possible. This chapter discussed the different types of services an airline might consider offering including: scheduled or non-scheduled; luxury, mid-range, low-cost, no-frills, shuttle, and charter; first, business, and economy class; food and beverage; in-flight entertainment; cargo and freight; in-flight duty-free; baggage restrictions; interline agreements; and other amenities the airline might offer to enhance its uniqueness. Under market analysis and opportunity, a number of questions were posed requiring answers in this section of the airline business plan. It was identified that this element's main goal is to provide the reader with a clear view of the market opportunity for the product and/or service. There was also a brief discussion on some of the factors requiring research concerning what airports to serve.

**Key Terms**

Scheduled service

Non-scheduled service

Luxury service

Mid-range service

Low-cost airline

No-frills airline

Shuttle service

Charter service

First class service

Business class service

Economy class service

Cargo and freight

Duty-free

Interline agreements

Hub-and-spoke airport system

Point-to-point airport system

Secondary airport

Slot system

## Review Questions

1.  What are the different types of airlines that exist?

2.  What type of airline is the business plan designed on?

3.  Does the airline business plan answer each of the questions posed in the discussion on market analysis and opportunity?

4. Given the different types of airports that exist, does the business plan target the right ones? What are the advantages and disadvantages to each of the airports selected?

**Web Sites**

http://www.netfeat.com/success/email_based_marketing.cfm

http://www.sallys-place.com/travel/business-travel/singapore_airlines.htm

http://www.airauto.com/aai/about/pressreleases/about_press_012201.htm

http://unpan1.un.org/intradoc/groups/public/documents/un/unpan001219.pdf

http://www.ttgasia.com/current/archive/2000/0421-27/co0420002102.html

**Recommended Reading**

Graham, Brian. *Geography and Air Transport.* Chichester, UK: John Wiley & Sons, 1995.

Radnoti, George. *Profit Strategies for Air Transportation.* New York, NY: McGraw Hill, 2002.

Shaw, Stephen. *Airline Marketing and Management, 4th Ed..* Burlington, VT: Ashgate Publishing, 1999.

Tretheway, Michael and Oum, Tae. *Airline Economics: Foundations for Strategy and Planning.* Vancouver, Canada: University of British Columbia, 1992.

Wells, Alexander T., and Wensveen, John G. *Air Transportation: A Management Perspective, 5th ed..* Belmont, CA: Thomson/Brooks-Cole, 2004.

# Chapter 10

## Analysis of Market Demand Levels

**Chapter Checklist**

- Purpose of forecasting

- Forecasting methods

- Market demand levels

**Introduction**

A great portion of the airline business plan's success depends on a thorough analysis of market demand levels through the use of expectations, predictions, and projections. Such an analysis is dependent on forecasting techniques. Forecasting is the attempt to quantify demand in a future time period. Quantification can be in terms of either dollars, such as revenue, or some physical volume, such as revenue passenger miles (RPMs) or passenger enplanements. Plans for the future cannot be made without forecasting demand. Planning also plays an important role in any aviation enterprise, but it should not be confused with forecasting. Forecasting is predicting, projecting, or estimating some future volume or financial situation – matters mostly outside of management's control. Planning, on the other hand, is concerned with setting objectives and goals with developing alternative courses of action to reach them – matters generally within management's control. A forecast of revenues is not a plan. There must be goals, strategies for attaining them, alternative courses of action, and a realistic fit with other market conditions. Thus, although forecasting is not planning, it is an indispensable part of planning, a management tool for deciding now what the company must do in order to realize its profit and other goals for the future. Not only is forecasting done for a given type of demand independently, but forecasts of one type of demand may also be based on other forecasts. Thus, the projection of flying hours for next year is an element in the forecast for future demand for flight personnel, fuel consumption, facilities, and a host of other considerations.

This chapter explains to the reader how to analyze market demand levels through the use of various types of data and travel characteristics, operating strategies of other airlines, and socio-economic characteristics of each potential service market. Also included is a discussion on statistical analysis methods, qualitative and quantitative research, and seasonality characteristics.

**Total Page Length: Approximately 2 Pages**

**The Purpose of Forecasting**

Each type of forecast serves a particular purpose. Thus, an airline might make a short-term forecast of total passenger enplanements between a particular pair of cities to provide a basis for determining station personnel and ground equipment needed, gate availability, and expenses related to these items. Short-term forecasts normally span a period of one month to one year and cover day-to-day operations as staffing stations, evaluating current competitive situations in the market, and projecting short-term equipment needs.

Medium-term forecasts generally span a period of one to five years and involve such things as route-planning decisions. A long-term forecast spans a period of five to ten years and might involve fleet planning decisions and long-term financial commitments. The various forecasts are used by companies to carry out three important management functions – analyzing, planning, and controlling.

*Analyzing*

Every airline, new or existing, must make choices among the many markets or submarkets open to it, in addition to deciding on the level of service to offer, the type of aircraft to fly on particular routes, and the type of aircraft to lease or purchase. The choice is greatly facilitated by quantitative estimates of demand.

*Planning*

Every airline, new or existing, must make short-term decisions about the allocation and scheduling of its limited resources over many competing uses; it must make long-term decisions about rates of expansión of capital equipment and funds. Both short-term and long-term decisions require quantitative estimates of demand.

*Controlling*

A company's actual performance (physical volume or revenues) in the market takes on meaning when it is compared to forecasts.

**Forecasting Methods**

The choice of forecasting methods should be based on several factors, including availability of data, accuracy of available data, management sophistication, intended forecast use, and availability of electronic data processing. Sophistication in forecasting methods can easily run ahead of data quality and management ability to use the results. Forecasting passenger enplanements for a one-year period on well-established routes, for example, poses a fundamentally different forecasting problem than estimating enplanements on a new route. Forecasting methods must be chosen accordingly. Annual forecasts are provided by various organizations, such as civil aviation authorities, International Air Transport Association (IATA),

International Civil Aviation Organization (ICAO), airport authorities, aircraft manufacturers, and others.

### *Causal Methods*

Causal (model) forecasts are based on a statistical relationship between the forecasted (dependent) variable and one or more explanatory (independent) variables. There need not be a cause-and-effect relationship between the dependent and independent variables. A statistical correlation alone is sufficient basis for prediction or forecasting. Correlation is a pattern or relationship between two or more variables. The closer the relationship, the greater the degree of correlation.

In general, a causal model is constructed by finding variables that explain, statistically, the changes in the variable to be forecast. Such variables must have the following characteristics: (1) they must be related statistically to the dependent variable, (2) data on them must be available, and (3) there must be some good way of forecasting them, or their relationship to the dependent variable must be lagged (must follow the dependent variable by several months).

Most forecasting methods are based on the assumption that existing patterns and historical relationships will continue in the future. Because this assumption usually holds only for the short term, however, most forecasting methods provide reasonably accurate forecasts for periods of only one or two years.

The statistical relationship is estimated and verified using statistical analysis. The selection of variables depends on the imagination and resources of the researcher. With the air of a computer, dozens of candidates can be tested, easily and quickly, once the structure – that is, the mathematical form – of the model has been decided. This, too, may be selected by trial and error.

The availability of data on the variables – or, more specifically, their specific values – is largely determined by the time and resources available to the researcher. Data is the key to specifying the model. Prominent independent variables used in forecasting various segments of the air transportation industry include gross national product (GNP), disposable person income (DI), and consumer spending on services. Dependent (forecast) variables might include such items as revenue passengers enplaned, RPMs, and passenger revenues. Other types of research include: history, curiosity and culture, business and professional, interpersonal, pleasure and romance, spiritual, sports, shopping, age, sex, education, social class, geographic region, city size, population density, family size, occupation, lifestyle, trip purpose, reliability, destination, availability, knowledge of alternates (airports), schedule convenience, cost, safety, unemployment rate, fixed investments, private housing trends, private commercial rate, government purchases, value of dollar to other currencies, fuel prices, domestic airline revenues, and domestic load factors.

Given unlimited amounts of data, causal models can be constructed that explain almost any market phenomenon. Unfortunately, unlimited amounts of data are rarely available. Shortages of time, money, and personnel; limits on the accessibility of data; deficiencies in measurement techniques – all impose serious constraints on data availability. Often, researchers must be content with secondary data, substitute variables, outdated observations, and inaccurate information. The result is usually an imperfect model, although not necessarily a useless one.

Forecastability, or a lagged relationship with the dependent variable, is essential, because it does little good to construct a forecasting model if the future values of the explanatory variables are as difficult to estimate as those of the dependent variable. The only alternative is to use independent variables whose present values determine the dependent variable's future values.

Causal models represent unquestionably the most sophisticated type of forecasting method used today, as well as the most frequently used one. However, it is important to be aware of their limitations when designing the airline business plan:

1. It is sometimes difficult to quantify all of the variables, even though the researcher is aware that these variables have influenced the dependent variable in the past and might continue to do so in the future.

2. It is often assumed that it is easier and more accurate to forecast explanatory variables (GNP, DI) than the dependent variable (passenger enplanements, cargo/ton-mile). This is important because the forecast variable is not better than the forecast of the independent variable.

3. It is often assumed that a functional relationship that existed in the past (and upon which the model was built) will exist during the forecast period.

### *Time-Series of Trend Analysis Methods*

Another reasonably sophisticated statistical method of forecasting is time-series analysis, the oldest, and in many cases still the most widely used method of forecasting air transportation demand. In some situations, this method is referred to as trend extension. It differs from causal model forecasting in that less causation is embodied in the time series.

Time-series models show the dependent variable as a function of a single independent variable, time. This method is used quite frequently when both time and data are limited, such as in forecasting a single variable for which historical data is obtained. Like the causal models, time-series models are based on a statistical correlation that does not necessarily reflect a cause-and-effect relationship between the dependent and the independent variable.

Aviation is certainly not static: new-aircraft sales, prices, revenue passenger miles, cargo tonnage, profits, flying hours, on-time performance, and number of departures fluctuate over time. Time-series or trend analysis is simply a sequence of values expressed at regular recurring periods of time. It is possible from these time-series studies to detect regular movements that are likely to recur and thus can be used as a means of predicting future events.

Forecasting through time-series or trend extension actually consists of interpreting the historical sequence and applying the interpretation to the immediate future. It assumes that the past rate of growth or change will continue. Historical data is plotted on a graph, and a trend line is established. Frequently, a straight line, following the trend line, is drawn for the future. However, if certain known factors indicate that the rate will increase in the future, the line may be curved upward, As a general rule, there may be several future projections, depending on the length of the historical period studied. Airlines keep numerous records of data of particular concern to them (departures, enplanements, flying hours, and so forth), and when a forecast is needed, a trend line is established and then projected out to some future time. The accuracy of forecasting by historical sequence in time-series or trend analysis depends on predictions of changing factors that may keep history from repeating itself.

The values for the forecast (dependent) variable are determined by four time-related factors: (1) long-term trends, such as market growth caused by increases in population; (2) cyclical variations, such as those caused by the business cycle; (3) seasonal phenomena, such as weather or holidays, and (4) irregular or unique phenomena, such as strikes, wars, and natural disasters. These four factors induce the following types of behavior in the dependent variable: (1) trends, (2) cyclical variations, (3) seasonal changes, and (4) irregular fluctuations. These types of variations are found throughout the literature of market and economic forecasting.

*Trends*

A trend is a long-term tendency to change with time. A variable's trend is a reflection of its statistical relationship with time, exclusive of cyclical, seasonal, and irregular disturbances. Trend functions are described by growth curves, which express, both graphically and mathematically, the underlying pattern of time-related changes. This pattern is usually brought about by such factors as population, GNP, industrialization, changes in technology, and long-term shifts in tastes or preferences. A trend can be inherently positive, such as total air carrier passenger revenues. It can be negative, such as the phasing out of fuel-inefficient aircraft from the airline fleet. Or it can be erratic, as in the case of airline pricing in recent years.

The time period specified for a particular trend varies considerably. Economists frequently define it as any period in excess of that required for a complete business cycle (approximately five years). Airline marketers tend to specify a trend period as equivalent to the approximate lifetime of the service. This can vary from a few months to a couple of years to an indefinite period.

*Cyclical Variations*

Cyclical variation is the variation of the forecast variable due to the business cycle. The business cycle is the wavelike fluctuation in the level of economic activity that has been associated with the economies of the developed nations since the early years of the Industrial Revolution. The business cycle has never been fully explained by economists, adequately controlled by governments, or satisfactorily predicted by businesses. However, the phenomenon is apparent if any of the common economic indicators (such as GNP, employment levels, stock prices, corporate profits, or capital investment) are plotted over time. The length of individual cycles varies, although they usually last well beyond a couple of years measured from peak to peak or valley to valley. In the United States, cycles range from 1 to 10 years, with 4 or 5 years being the norm. The magnitude of the fluctuations, measured vertically from peak to valley (or vice versa) varies tremendously and thus far has defied precise forecasting, to the chagrin of most aviation industry analysts.

The business cycle has a significant effect on all the segments of the air transportation industry. The level of air travel for business or pleasure purposes is affected by upturns and downturns in the economy. Economists refer to the air transportation industry as being income elastic; that is, airplane sales, RPMs, and so forth are very responsive to changes in economic aggregates such as disposable income, personal income, and national income.

*Seasonal Variations*

Seasonal variation is the variation of the forecast variable associated with the time of year. It is appropriately named, for it is a function of both the weather and the social customs associated with the four seasons. Seasonal fluctuations, peaks and troughs, in the demand for such things as hotel rooms, rental cars, and airline travel are quite pronounced.

*Irregular Variations*

Irregular variations are erratic, non-recurrent events such as strikes, blizzards, riots, fires, wars or war scares, epidemics, price wars, bankruptcies, and other real-world disturbances. Although the disturbances factor is easily identified and the magnitude of its effect can normally be estimated, it seldom can be forecast. Events to hit the airline industry in early part of the twenty-first Century include the terrorist attacks of 9/11, the Iraqi crisis, SARS, and continued threats by terrorists. Although the disturbance factor is easily identified and the magnitude of its effect can be estimated, it seldom can be forecast.

Consulting mainstream references on forecasting will identify different strategies that can be incorporated into the airline business plan when looking at cyclical, seasonal, and irregular variations.

**Judgmental Methods**

Judgmental forecasts are educated guesses based on intuition and subjective evaluations. Although they are the least rigorous types of forecasts, they are frequently used as a powerful factor in decision making. Intuition often is the only tool the researcher has, and it can be very accurate. Judgmental methods can be used when either no information or very little historical data exists. They can also be used to adjust forecasts developed by causal models or through time-series analysis.

Forms of judgmental methods include expert opinion and poll forecasts. The usefulness of these methods depends on the cost, availability, and reliability of these types of data.

**Market Demand Levels**

The airline business plan must incorporate a comprehensive analysis of potential demand in the markets served. Although sometimes difficult to obtain, the most accurate available data concerning passenger travel characteristics, operating strategies for other airlines, and socioeconomic characteristics of each potential service market must be used. Generally, information obtained from civil aviation authorities, major airlines, regional and local chamber of commerce, and economic development agencies is considered a reliable source. However, be careful when consulting foreign sources because different countries and regions of the world have different reporting methods meaning that the data obtained might not be reliable.

Based on your forecast analysis, this section of the business plan should explain to the reader the following conclusion:

*In summary, X Airways market demand models reveal that the airline will reach and maintain the following levels of demand for air service into X airport from its outstation markets in the long term. The following values represent an average weekly demand volume over a one year period beginning calendar year 20XX.*

| City A | 290 (Jan-Jun) | 125 (Jul-Dec) |
|---|---|---|
| City B | 185 (Jan-Jun) | 92 (Jul-Dec) |
| City C | 1,420 (Jan-Jun) 985 (Jul-Dec) | |
| City D | 750 (Jan-Jun) | 640 (Jul-Dec) |
| City E | 350 (Jul-Dec) | 290 (Jan-Jun) |
| City F | 290 (Jul-Dec) | 173 (Jan-Jun) |
| City G | 230 (Jul-Dec) | 155 (Jan-Jun) |

Based on the above example, this section of the business plan should state the increase in rate of demand as an annual percentage as follows:

*Should forecast demand levels increase as predicted, X Airways is poised to expand its service fleet and market frequencies.*

Although not a very lengthy portion of the airline business plan, this section should also discuss the competitor(s) and potential competitor(s). Responses should be customer-focused because ultimately, knowledge of the passenger's needs and demands is what makes a successful airline. For each competitor, the following information is needed: sales volume, market share, market strategies, competitive style, price, performance, and a thorough SWOT (strengths, weaknesses, opportunities, threats) analysis. The business plan might also incorporate a PESTE analysis for the study of the airline's marketing environment. PESTE stands for Political, Economic, Social, Technological, and Environmental factors. Depending on the structure of the business plan, this portion might be included under the title of *Analysis of Market Demand Levels* or *Sales and Promotion Strategy* (see *Chapter 13*).

**Summary**

This chapter discussed how to analyze market demand levels through the use of different forecasting techniques including: causal methods, time-series of trend analysis methods, and judgmental methods. Forecasting is used to carry out three important management functions – analysis, planning, and control. Such techniques can be used to measure quantification using a number of different terms. It is important to note that forecasting techniques should not be used solely on their own as they must be combined with integrated planning techniques. A forecast is not a plan but merely a tool to be used with planning alternatives. Goals, strategies, alternative courses of action, and a realistic fit with market conditions are necessary for effective results when forecasting.

**Key Terms**

Forecasting

Causal (model) forecast

Correlation

Time-series analysis

Trend extension

Judgmental forecast

SWOT analysis

PESTE analysis

**Review Questions**

1. How does forecasting differ from planning? What is the purpose of forecasting?

2. Describe how forecasts can be used by new and existing airlines for analysis, planning and control?

3.  What are the different types of forecasting models available for development of the airline business plan? What are the strengths and weakness of each?

4.  What are the differences between short-term and long-term forecasts? What are the advantages and disadvantages of each?

## Web Sites

http://www.AirlineMonitorWeekly.com

http://www.ny.frb.org

http://www.rati.com

## Recommended Reading

Armstrong, J. Scott. *Principles of Forecasting*. Norwell, Mass.: Kluwer Academic, 2001.

Box, George E. P., and Gwilyn M. Jenkins. *Time Series Analysis: Forecasting and Control*. San Francisco; Holden-Day, 1970.

Brown, Robert G. *Smoothing, Forecasting, and Prediction of Discrete Time-Series*. Englewood Cliffs, N.J.: Prentice-Hall, 1963.

Chisholm, Roger K., and Gilbert R. Whittaker, Jr. *Forecasting Methods*. Homewood, Ill.: Irwin, 1971.

Makridakis, Spyros. *Forecasting, Planning, and Strategy for the 21$^{st}$ Century*. New York: Free Press, 1990.

Mentzer, John T., and Carol C. Bienstock. *Sales Forecasting Management: Understanding the Techniques, Systems, and Management of the Sales Forecasting Process*. Thousand Oaks, Cal: Sage 1998.

Shaw, Stephen. *Airline Marketing and Management, 4$^{th}$ ed.* Aldershot, England: Ashgate, 1999.

Taneja, Naval K. *Airline Traffic Forecasting*. Lexington, Mass.: Heath, 1978.

Wells, Alexander T., and Wensveen, John G. *Air Transportation: A Management Perspective, 5ʰ ed*. Belmont, CA: Thomson – Brooks/Cole™ , 2004.

# Chapter 11

## Proposed Route Structure and Schedule

**Chapter Checklist**

- Purpose of scheduling

- Maintaining of aircraft equipment

- Flight operations and scheduling of crews

- Ground operations

- Schedule planning and aircraft assignment

- Guidelines to follow

## Introduction

This element of the airline business plan focuses on the proposed route structure and schedule. Schedules represent one of the primary products of an airline and certainly one of the leading factors in a passenger's choice of a particular carrier. Scheduling is a very difficult task because of the number of factors involved and is as important as forecasting, pricing, fleet planning, or financing. Much of the airline's success is dependent on an efficient schedule. This chapter introduces the reader to the scheduling concept, equipment maintenance, flight operations and crew scheduling, schedule planning, and aircraft assignment. The chapter concludes with specific guidelines to follow when designing this element of the business plan.

## Total Page Length: Approximately 3 Pages

## Purpose of Scheduling

An efficient airline schedule performs four main tasks. Firstly, the schedule provides adequate service based on passenger demand. Secondly, provides economic strength for the company in terms of profitability. Thirdly, the airline schedule provides for sales and competitive effectiveness. Fourthly, it provides operational dependability and efficiency.

When designing the schedule, various factors must be incorporated. Major internal factors include: equipment maintenance requirements, flight operations (airport runway lengths, fuel capacity, air traffic control and routings, crew availability), airport facility constraints (gate positions, ticket-counter space, baggage handling, ground equipment, food service), and marketing factors (traffic flow, sensitivity of schedule salability, other operating factors, load factors). Major external factors include: airport authorities (curfews, slots, other restrictions), local communities, hotel and cruise ship operators, travel agents, tour operators, and air freight shippers.

Airline scheduling impacts all aspects of an airline operation working with all the departments and field stations. Regardless of the airline's size, there should be an established scheduling department headed by a chief architect or "choreographer". It is important for this department to work closely with marketing administration because of the overriding importance of service to the public.

## Maintaining of Aircraft Equipment

There are different options available for the maintaining of aircraft equipment. Depending on the size of the airline, availability of funds, and whether the airline leases or owns its aircraft, the airline business plan should identify the most efficient method suitable for the operation. Safety should be the prime concern for any airline meaning that safety should override economics. However, some airlines have been found guilty of neglecting maintenance issues in order to save a few dollars. This is particularly true with some charter and low-cost carriers but such operations are mainly concentrated in developing countries where regulations are not as strict as in the developed countries. At the end of the day, this is not a smart move. Different maintenance options include in-house maintenance or subcontracted maintenance. If an airline selects to perform maintenance in-house, a large amount of physical infrastructure is required as well as key personnel to perform maintenance tasks. For new airlines starting out, this is a very expensive option. If an airline decides to subcontract maintenance, aircraft can be sent to a major airline's maintenance facility or to a licensed maintenance repair overhaul facility. For new airlines launching service, this is most likely the best option to select. Some airline industry experts argue that if an airline is planning to operate less then five aircraft, maintenance should be subcontracted. If this option is chosen, it is important to research the different facilities available and be sure to check background history. Some facilities are better than others and when safety is the prime focus, it might be worth spending extra dollars. A safe operation should be based on four maintenance efficiency goals: (1) minimize aircraft out-of-service time, (2) use up time allowable on aircraft parts between overhauls, (3) seek optimum utilization of personnel and even workload, and (4) maximize utilization of facilities due to substantial investment in buildings, tooling, and specialized equipment. It is important to select a facility that shares the same maintenance efficiency goals.

## Flight Operations and Scheduling of Crews

Depending on the type of airline being designed, there are different options available for the scheduling of flight crews. Some airlines will hire and train their own staff while others might subcontract this aspect of the business to an aircraft operator or to a company providing qualified staff trained according to the type of aircraft flown. Regardless of what option is selected, there are a number of operational factors that must be incorporated into scheduling. These include: airport runway lengths, aircraft fuel capacity, adverse weather, air traffic control and routings, crew time limits, and labor issues. Depending on what region of the world the airline is operating in, it is important for the airline business plan to identify working limitations that govern flight crews so that limitations are not exceeded. Working limitations are normally set by a country's civil aviation authority and the airline's company agreement. For example, in the United States, the Federal Aviation Administration (FAA) permits a daily limitation of 16 hours maximum flight duty time for pilots on a two-person crew, unless, prior to exceeding 16 hours, a rest period was provided of no less than 10 hours. An increase of a few minutes to the flight schedule could result in a forced crew break or layover that could have been avoided. Such a delay could mean increased costs and inefficiency for the airline. Additionally, the FAA restricts flight crews to a minimum of 8 hours of rest in any 24-hour period that includes flight time. Flight crews cannot exceed 40 hours during any seven consecutive days.

## Ground Operations

When it comes to ground handling, there are different options available for the airline and the business plan should identify the preferred method(s). One option includes performing all the ground handling duties in-house. Such an option requires high investment into physical infrastructure and personnel. The second option includes subcontracting ground handling duties to another airline or ground handling company. The first option is normally selected by major airlines although at some destinations, the subcontract option is often used. For a new airline starting out, subcontracting of ground handling makes the most sense because it is economical, compared to the high cost of acquiring infrastructure. This is especially true for airlines offering seasonal flights to destinations or limited frequency It should be noted that there are two types of ground handling categories. The first is defined as "above the wing" while the second is defined as "below the wing". The first category handles areas like check-in and baggage handling while the second category handles area like fueling, catering, and cleaning.

When designing the route schedule, the following factors should be considered:

1. Are there enough gate positions for the number and type of aircraft on the ground simultaneously, including early arrivals and delayed departures?

2. Is there adequate ticket-counter space to handle passengers quickly and efficiently?

3. Is sufficient time provided for on-line or interline transfer of passengers, baggage, mail, cargo and freight?

4. Can the planned flights be handled efficiently by counter, ramp, and catering personnel? If not, will additional revenue from new flights or a new connection be sufficient to more than offset the cost of additional personnel?

5. Will the proposed route schedules introduce a second or a third personnel shift? Will a minor flight adjustment permit the reduction of one shift?

6. Is there ground equipment of the right type: aircraft starter units, baggage vehicles, cargo conveyors, forklifts, tow tractors, air stairs? If not, how will the proper equipment be acquired and at what cost? Can such equipment be borrowed from another carrier or ground handling company?

## Schedule Planning and Aircraft Assignment

When determining the size of a given market, projecting future growth, estimating the effect of planned product changes on the size of the total market and on the airline's own share of the market, attempting to forecast moves by the competition, or estimating the costs and revenues of alternative plans of action, a number of factors should be identified since they will impact the assignment of aircraft on the proposed schedule. These factors include: traffic flow (number of originating and connecting passengers on a given route), schedule salability (days and times demanded by passengers), time zones (curfews, departure and arrival times), station personnel (minimize peaking of equipment and ground personnel), equipment turnaround time (amount of time required to service the aircraft), chain reaction effect (schedule changes due to delays), and load factor leverage (the percentage of seats occupied by revenue passengers). Additional information related to these factors can be found in the recommended reading section of this chapter.

When designing the route schedule, this element of the airline business plan should decide upon the type of schedule it plans to operate. There are four basic schedule types: (1) skip-stop, (2) local service, (3) cross-connections (hub and spoke), and (4) non-stops. A skip-stop schedule provides service to points A, B, C, D, E, F, G, and so forth by scheduling flights in the following manner: A-C-D-G or A-D-G, or similar combinations in which one or more of the intermediate flights are "skipped" with service being offered by other flights. This type of schedule provides fast service to intermediate stations but does not provide service between consecutive cities. A local service schedule is beneficial for short-haul flights making all stops on a segment and connecting at larger intermediate stations with larger aircraft. This type of schedule provides fast service between small intermediate stations and terminal points but a change of aircraft if required. A cross-connection (hub and spoke) schedule provides services that maximizes connectivity between points in a short period of time (usually 45 minutes or less). A non-stop schedule simply offers service between two points. In today's aviation environment, small airlines find it beneficial to offer a point-to-point service usually utilizing secondary airport facilities.

## Guidelines to Follow

The proposed route structure and schedule should be based on physical requirements of travel between markets, efficient by serving the most markets with the least number of aircraft required flight hours, and cost-effective by matching service levels with demand. This element of the airline business plan should explain how the time schedule was created (For example, factors demand levels are based on) followed by a brief explanation of operational strategy (For example, increasing service frequencies to proposed markets and expansion to additional markets). The proposed route structure should use standard coding followed by the Official Airline Guide (OAG). Departure and arrival times should include local and GMT. Departure and arrival cities should be identified using the official three-letter designator. For example, London Heathrow's designator is LHR. The schedule should also include the suggested one-way fare (based on detailed demand vs. cost analysis).

The first paragraph of this section on proposed route structure and schedule might resemble the example below:

*X Airways has developed an initial route structure that capitalizes on the opportunity to serve the markets described in a manner that is feasible, based on physical requirements of travel between markets, efficient, by serving the most markets with the least number of required flight hours, and cost-effective, by matching service levels with forecast demand.*

This statement might be followed by:

*The X Airways proposed route structure is detailed below. Frequency is described by the day(s) of the week service is provided, using standard coding followed by the Official Airline Guide (OAG).*

*1 = Monday, 2 = Tuesday, 3 = Wednesday, 4 = Thursday, 5 = Friday, 6 = Saturday, 7 = Sunday*

Following the above two paragraphs, the actual time schedule should be listed by city-pair.

*All times are in local and GMT time.*

*Service between Orlando International Airport (MCO) and:*

*Newburgh (SWF), Five Times Weekly Service*

| Departs | Arrives | Day Served | Dep. Time | Arrival Time |
|---------|---------|------------|-----------|--------------|
| *MCO* | *SWF* | *1,2,3,4* | *0600 (1100)* | *0806 (1306)* |
| *SWF* | *MCO* | *1,2,3,4* | *0900 (1400)* | *1100 (1600)* |
| *MCO* | *SWF* | *5* | *0500 (1100)* | *0706 (1206)* |
| *SWF* | *MCO* | *5* | *0800 (1300)* | *1006 (1506)* |

*Suggested One-Way Fare: $92.50*

-    *Insert all destinations here –*

The next section of the schedule should include the following:

*This schedule has been designed to* (list strategies point by point as to how the schedule was designed):

    *a) Strategy*
    *b) Strategy*
    *c) Strategy*
    *d) Strategy*
    *e) Strategy*

This element of the airline business plan should conclude with approximately three paragraphs outlining the calendar operating period, the airline's operational strategy, and how the suggested one-way fares were calculated.

## Summary

This chapter concentrated on one element of the airline business plan – the proposed route structure and schedule. The main subjects discussed included: the scheduling concept, equipment maintenance, flight operations and crew scheduling, schedule planning, and aircraft assignment. Scheduling is a complex process consisting of numerous variables related to all aspects of airline operations. In short, the schedule is the nucleus of the airline business plan. Regardless of the airline's size or corporate structure, all departments must interact with one another and contribute to the establishment of an successful schedule that performs four main tasks: provides adequate service based on passenger demand, provides economic strength for the company in terms of profitability, provides for sales and competitive effectiveness, and provides operational dependability and efficiency.

**Key Terms**

Profitability

Competitive effectiveness

Operational dependability

Schedule salability

Above-the-wing services

Below-the-wing services

Traffic flow

Chain reaction effect

Skip-stop

Cross-connections

Official Airline Guide (OAG)

**Review Questions**

1.  What is the main purpose of scheduling and what are the four main tasks scheduling should accomplish?

2.  What are the main internal and external factors that should be incorporated when designing the schedule?

3.  What are the different factors that must be considered when designing the aircraft maintenance program? What are the different options available for the airline in terms of the different types of maintenance programs available?

4.  What questions should be answered in relation to the establishment of ground handling?

5.  What are the different types of route schedules that exist? What option is best for the airline proposed?

**Web Sites**

http://www.airlineguide.info

http:///www.oag.com

http://www.omegaair.com

**Recommended Reading**

Radnoti, George.  *Profit Strategies for Air Transportation*.  New York, NY: McGraw Hill, 2002.

Shaw, Stephen. *Airline Marketing and Management, 4th Ed.*.  Burlington, VT: Ashgate Publishing, 1999.

Tretheway, Michael and Oum, Tae.  *Airline Economics: Foundations for Strategy and Planning*.  Vancouver, Canada: University of British Columbia, 1992.

Wells, Alexander T., and Wensveen, John G. *Air Transportation: A Management Perspective, 5th ed.*. Belmont, CA: Thomson – Brooks/Cole™ , 2004.

# Chapter 12

## Financial Analysis

**Chapter Checklist**

- Income statement (expenses)

- Income statement (revenue)

- Operation plan

- Summary financial projections for 5-year business plan

- Statistical summary for a 5-year period

- Return on investment

## Introduction

This chapter introduces the reader to perhaps the most important section of the airline business plan as it provides the reader with a set of financial projections that show the financial implications of the business plan. Make sure that you clearly state the assumptions behind these projections, both in terms of costs and revenues, and ensure that they are realistic. By explaining the assumptions clearly, a potential investor will be able to see that the figures have been thought through - whether or not he/she agrees with them. This section of the business plan must provide answers to the lender's questions as follows: When will we start making money on this venture? How much will we make if we meet projections? What about if we do not meet projections? When and how will we be able to pay off any loans required?

The financial projections should include the following items:

- Income statement (expenses)

- Income statement (revenue)

- Operation plan

- Summary financial projections for 5-year business plan

- Statistical summary for a period of 5 years

**Total Page Length: Approximately 8 to 10 Pages**

## Income Statement (Expenses)

This section of the financial analysis of the business plan must provide the reader with an absolutely complete set of information on expenses made by the new airline. A special attention must be paid to the fact that the airline industry is extremely specific and quite different from other types of common industries, therefore, the business plan itself and especially its financial analysis will be very specific as well, and commonly used business knowledge might not be quite sufficient.

The income statement portion dealing with expenses should be produced for the first four quarters of operation and for the first five years of operation. It should contain the following categories of individual expenses:

*Airplane-Related Costs*

- Cockpit/cabin crew
- Fuel and oil (see below)
- Maintenance
- Aircraft service
- Landing fees
- Navigation fees

*Traffic-Related Costs*

- Traffic servicing, reservation, and sales
- Food, liability insurance

*System-Related Costs*

- Advertising and publicity ($ for 6 months, X% of revenues later)
- General and administrative
- Taxes

*Annual-Related Expenses:*

- Depreciation (amortization)
- Hull insurance
- Lease/rental
- Finance (interest) charges

*Flying operations (cockpit crew):*

- Gross weight pay (type of aircraft)
- Mileage pay
- Benefit pay
- Training pay
- Personal expenses (hotels, meals)
- Payroll taxes

**Fuel costs**

Fuel costs represent the highest costs of any airline's operation, therefore, the highest possible attention should be paid not only to acquisition of fuel at optimum prices but also to precise calculations of the amounts of needed fuel. Those calculations may be based on the following factors:

- Derated takeoff (lower engine temperature, extends engine life, lowers
- maintenance costs)
- Climb schedule
- Cruise policy (Mach)
- Descent schedule
- Optimal altitude
- Reclearance policy
- Tankering
- Fuel flow degradation based on aircraft age
- Conduct performance audit
- Airport characteristics and payload

*Insurance*

- Hull (certain amount of the aircraft price less engine costs)
- War-risk (3 to 5% of aircraft hull cost)
- Passengers (based on mileage; 10 to 30 cents per passenger)

*Other costs*

- Financing cost (buildings, sales, offices, hangars, maintenance facilities, computers, communication equipment)
- Landing fees and navigation fees
- Station costs (flight processing, weighing, supervision, transit in and ramp building, passenger handling, cargo handling)
- Systems overhead (general and administrative, advertising and publicity, HQ building and office maintenance, salaries and wages of personnel not yet classified)

*Start-up costs*

- Facilities
- Spares
- Ground equipment
- Training (flight crews, cabin crews, maintenance, other personnel, and new hires)

Of course, the question is where to obtain all that information and how to make sure it is correct. Suitable sources of information include real-life costs and projected revenues, FAA reported data (Database Products, Inc), Form 41 and O&D, Conklin and Decker Associates, Inc., Airport authorities, Flight Safety Boeing (training costs), Flight Safety International (Atlanta), (initial and recurrent training costs – pilots – F/As), Brown and Brown (insurance), and aircraft leasing companies (ILFC, GPA Group).

**Income Statement (Revenue)**

An airline can generate its revenues through the following channels:

- Passengers
- Cargo
- Excess baggage
- Mail
- Miscellaneous revenues from other sources

*Passengers*

- Yield – air transport revenue is expressed per unit of traffic carried
- Shown in cents/passenger mile or cents/passenger kilometer
- It must be known for each city-pair and the entire system
- Generally applied to all revenues obtained from scheduled revenue-passenger miles flown, including non-revenue passengers

*Freight (if applicable)*

- Increasingly more important in airline revenues
- Demand greater than passengers
- Revenue is measured in cents per ton-mile or cents per ton-kilometer
- Mail revenue is measured in the same way

*Excess Baggage*

- It is very significant in operation of international airlines
- It is expressed in percentages of passenger revenues

*Miscellaneous revenues from other sources (if applicable)*

- Charter operations
- Royalties
- Ground transportation
- Handling services

- Rents
- In-flight sales
- Contract maintenance
- Training
- Other revenue-generating services

Passenger revenue is a function of various fare levels offered to the public on different routes and the total number of passengers carried. It is recommended to propose one-way fares (excluding Federal taxes). It must be also considered whether other transport revenues (freight, mail) will be included or not.

## Operation Plan

The operation plan is usually a table comprising the following sectors:

- Scheduled weekly round trips by sector
  FROM/TO/AIRCRAFT/YEAR1/2/3/4/5
- Scheduled weekly departures by station
  FROM/TO/AIRCRAFT/YEAR1/2/3/4/5
- Aircraft fleet (number of aircraft for each year)
- Utilization hours per day

The operation plan represents the foundations for the financial projections.

## Summary Financial Projections for 5-Year Business Plan

As soon as the operation plan is produced, one can start forming financial projections for a five-year plan, usually in a form of a one-page table. The purpose of that summary is to provide the reader/investor with an opportunity to see the expected financial developments in a nutshell. The financial projections for 5-year business plan comprise the following items:

- **Operating revenues**
- **Operating expenses**
- **Operating income/(loss)**
- **Net income/(loss) before taxes**
- **Operating margin**
- **Net margin before taxes**

## Statistical Summary for a 5-Year period

The statistical summary in the form of a table comprises the following items:

- Revenue passenger carried
- Revenue passenger miles (RPMs)
- Available seat miles (ASMs)
- Passenger load factor
- Break even load factor
- Passenger revenue per ASM
- Operating expense per ASM
- Total expenses per ASM
- Stations in operation
- Departures
- Block hours
- Aircraft miles flown
- Total seats
- Average seats/aircraft
- Number of aircraft
- Utilization (hours per day)
- Average stage length
- Average block time
- Fuel cost per gallon (average)

## Return on Investment (ROI)

Should the reader read all of the above mentioned information, he/she will then want to know when his/her investment will return. In other words, the business plan must quite clearly specify the return on investment (ROI). One of the most frequently used methods rests in simple charts that are easy to produce and understand.

The following charts are usually included in the business plan:

- Profit margin by X airways markets
- 20 quarters (5 years)
- Profit (loss) per quarter
- 20 quarters (5 years)
- Projected result of operation
- 20 quarters (5 years)

The most costly piece of equipment is the aircraft, therefore, the ROI calculation must be based on its (their) delivery date(s), economic life, purchase price, depreciable price, direct operating costs and indirect operating costs, depreciation method, income tax rate, residual value, aircraft residual value at the end of its economic life, pre-delivery payment schedule, and on miscellaneous additional information. One of the very critical pieces rests in timing (all cash flow must be strictly time accountable because of time value of money) and the level of interests.

## Summary

The goal of this chapter was to introduce the reader to the most important section of the airline business plan because there are a lot of questions and doubts in every lender's mind which need to be appropriately answered if a venture capital is to be obtained. Information provided to the lender should be, therefore, the result of serious and thorough research and correct calculations based on available actual data. During production of financial projections it is always better to be realistic and to reach modest profits than to waste huge funds later as a result of exaggerated assumptions.

The airline business plan must contain the following: income statement (expenses), income statement (revenue), operation plan, summary financial projections for 5-year business plan, and statistical summary for a period of 5 years.

## Sample Business Plan

In order to facilitate and expedite production of a real airline business plan, a sample airline business plan produced by the author of this book can be used. Its business concept is strictly hypothetical and is based on a fictitious airline. The sample business plan does, among other things, an excellent job of using Excel™ spreadsheets to break down all the financial details of running an airline. It is available on a CD-ROM and can be obtained directly from the author for a nominal fee.

## Key Terms

Financial projections

Income statement

Operation plan

Summary financial projections

Statistical summary

Fuel costs

Return on investment

## Review Questions

1. What are the components of the financial analysis of a business plan?

2. What is the purpose of the financial analysis?

3. Why do we use the income statement?

4. What items comprise the income statement (expenses)?

5. What items comprise the income statement (revenue)?

6. What is the purpose of the statistical summary?

7. What is the return on investment and what is it based on?

## Web Sites

http://sbinfocanada.about.com/library/weekly/aa090503a.htm

http://www.onlinewbc.gov/docs/starting/bp_essentials.html#financials

http://www.bcentral.co.uk/finance/plan/FinancialForecasting.asp

http://www.planigent.com/html/template.html

## Recommended Reading

Jablonsky, Stephen F. and Barsky, Noah P. *The Manager's Guide to Financial Statement Analysis, 2nd edition.* New York, NY: Wiley, 2001.

Miller, Barry E. and Miller, Donald E. *How to Interpret Financial Statements for Better Business Decisions*. New York, NY: AMACOM, 1991.

Miller, Barry E. and Miller, Donald E. *Pushing the Numbers in Marketing: A Real World Guide to Essential Financial Analysis.* Westport, CN: Quorum Books, 1992.

Rados, David L. *Pushing the Numbers in Marketing: A Real World Guide to Essential Financial Analysis.* Westport, CN: Quorum Books, 1992.

# Chapter 13

## Sales and Promotion Strategy

**Chapter Checklist**

- The marketing mix

- Entry strategy

- How do we get passengers?

## Introduction

Writing a successful airline business plan does not mean the airline will be successful. Once the foundation has been laid, marketing the carrier's product and/or service becomes extremely important. Marketing directs the flow of services provided by the airline to the customers' needs and wants to achieve the company objectives. The subject of marketing goes beyond simply selling and incorporates numerous business activities including forecasting, market research and analysis, product research and development, price setting, promotion, and finance activities such as credit and collection associated with ticket sales. This element of the airline business plan describes how the airline will attract passengers to the carrier through the use of different marketing strategies. This chapter concentrates on the main factors that should be researched when developing an entry strategy to promote business.

### Total Page Length: Approximately 1 to 3 Pages

## The Marketing Mix

In order for the airline to reach its target market, a number of marketing activities are necessary. These marketing activities are referred to as the marketing mix. The marketing mix consists of the types and amounts of controllable marketing-decision variables that a company uses over a specific time period. Commonly referred to as the "four Ps", these variables are:

1. Product – The right product (or service) must be developed for the target market.

2. Price – A price that gives good value to the customer and adequate revenue to the carrier must be set for the product.

3. Promotion – Personal selling and advertising must be used, both to communicate information about the product to the customer and to facilitate sales.

4. Place – Appropriate channels of distribution must be found to ensure that the product reaches the target market at the right time and in the right place.

Before this element of the airline business plan can be designed, it is important to have an understanding of the uncontrollable variables associated with the marketing mix. These variables are:

1.  Cultural and social differences – These are the traditions and values of various ethnic groups that represent potential customers.

2.  Political and regulatory environment – Political climates are constantly changing. New levels of taxation and government spending can affect marketing strategies set by the carriers. Regulatory requirements, such as allocations of landing quotas at certain airports because of extreme peaking in the number of flights, can undermine the best of marketing plans.

3.  Economic environment – A good marketing program might fail if the economy is experiencing a recession or rapid business downturn. Airlines are very sensitive to changes in the economy creating the need for flexibility within the airline business plan.

4.  Existing competitive structure – The number and types of competitors the marketing team must face in its target markets may vary.

5.  Resources and objectives of the company – Top management really controls these variables, and the marketing team members must work within the restraints imposed on them.

    Also, in marketing the airline product, there are certain unique characteristics that must be recognized:

1.  The product (service) cannot be kept in inventory to match fluctuations in demand. The revenue lost as a result of an unfilled seat when the aircraft door closes is lost forever.

2.  The service is usually personalized. Two people who take the same flight might come away with completely different opinions about the service, depending on their individual experiences.

3.  There is no such thing as replacement of a bad product, as is the case in the sale of other products.

4.  It is difficult to check the quality of the service before the final sale. There is no showroom to visit to test the product before purchase.

5.  Delivery of the product cannot always be guaranteed, due to mechanical problems or the unpredictability of the weather.

6.  The service can be produced only in batches, as opposed to individual units.

**Entry Strategy**

The sales and promotion strategy portion of the airline business plan should incorporate an entry strategy to give the reader a clear view of how management proposes to get the business started. It answers the question, "What do we have to do to be in the marketplace? How are we going to get our initial paying customers?"

When the product and/or service is ready to be sold, the business plan should describe how it will be introduced to the market: in the fastest way possible; at the lowest risk, should it fail; at the lowest cost; to preempt competition and to gain targeted initial market share quickly.

The entry strategy should describe the airline's marketing, advertising, and promotional strategies during this time period: Product positioning statements and validation testing; Where and how do your prospects hear about your product?; What will they hear? When? "The message?"; How you will achieve the communication objectives, by market segment of arousing awareness, interest, need and/or want.

Which of the following will the airline use during the introductory phase? For those used, how will the sub-issues be addressed: Advertising? (print, radio, TV, geographic targets, frequency, pre-testing program, competitive advertising); Direct Mail Communications? (obtain lists, fulfillment resources, response rates expected, how leads will be handled); Telephone Marketing? (who, training required); Promotional Events? (trade shows, seminars and/or workshops); Publicity and Public Relations? (news releases, articles, speeches, industry associations); Sales Promotions? (On which customers within which target segments will you focus?).

**How Do We Get Passengers?**

Once the main factors of entry strategy are understood, the developer of the airline business plan should find the task of attracting passengers to the airline easier. Aside from the mainstream marketing strategies introduced earlier in this chapter, there are additional factors to consider that are more directly linked to the airline business. Airlines have the option of utilizing travel agents, tour operators, computer reservation systems (CRSs) or global distribution systems (GDSs). The use of the travel agent has declined in some parts of the world because the travel agent is paid a fee for each ticket sold. Airlines have realized how much money is spent on travel agent commissions and have promoted other forms of selling tickets to keep costs down ultimately passing the savings on to the passenger. Tour operators are beneficial to airlines offering tour packages or charter carriers wanting to pass on the pressure of ticket sales to another party. Tour operators often purchase a high volume of seats on board a particular flight and accept responsibility of selling the booked seats to travel agents, other tour operators, or to the public. Subscribing to a CRS or GDS can be beneficial because flight details are logged into a computer program permitting access to travel agents that ordinarily might not be familiar with your airline. However, subscribing to such a system can be expensive and it is important to note that many of the existing reservation systems are owned by major airlines.

The preferred method to attract passengers to the airline is through direct selling methods like the Internet. The Internet eliminates the middleman keeping costs low for the airline and creating savings for the passenger. It also allows a 24-hour access, 365 days per year enticing potential customers. However, for a new airline starting out, the use of the Internet can be difficult because the brand name is not recognized. In today's aviation environment, new airlines starting out might find it easier to utilize travel agents and tour operators until the brand name becomes established. If connectivity is not an issue and there are no established interline agreements, it is recommended that the airline not subscribe to one of the main CRSs. A customized system can be designed for an individual airline at a much lower cost.

Other methods to attract passengers include frequent flier programs (FFPs), code sharing with other airlines, and incentives for passengers to fly with a particular airline. Many airlines also offer electronic tickets and electronic and remote check-in to reduce costs and make the air travel experience more convenient for the passenger.

## Summary

This chapter discussed sales and promotion strategy as one element of the airline business plan. It was mentioned that if a company has a sound business plan, it does not mean the company will be necessarily successful. Success is very much dependent on marketing strategies that attract passengers to the airline. Before discussing any of those strategies, this chapter introduced the reader to the marketing mix and introduced controllable and uncontrollable marketing-decision variables. A full understanding of these variables is necessary before the developer of the airline business plan can establish sales and promotion strategies that will be successful for the type of airline being launched. This element of the business plan should incorporate an entry strategy and the main factors associated with this subject were highlighted followed by a brief discussion on additional techniques used to attract passengers to the airline.

**Key Terms**

Marketing Mix

Product

Price

Promotion

Place

Entry Strategy

Computer Reservation System (CRS)

Global Distribution System (GDS)

**Review Questions**

1. What is meant by the marketing mix and what are the 4 Ps?

2. What are the controllable and uncontrollable marketing-decision variables that must be considered when developing the airline business plan?

3. What unique characteristics must be acknowledged in relation to the airline product before the business plan is completed?

4. What is meant by the entry strategy and what factors should be addressed within the sales and promotion strategy portion of the airline business plan?

5. What are some of the strategies used by airlines to attract passengers to the carrier?

**Web Sites**

http://www.expresstravelandtourism.com/20030331/airwaves4.shtml

http://phoenix.bizjournals.com/phoenix/stories/2003/08/11/daily6.html

http://www.executivelibrary.com/Library.asp

http://sbinformation.about.com/library/weekly/aa111002a.htm

## Recommended Reading

Abeyratne, Ruwantissa. *Aviation Trends in the New Millennium.* Burlington, VT: Ashgate Publishing, 2004.

Dempsey, Paul Stephen and Gesell, Laurence. *Airline Management: Strategies for the 21st Century.* Chandler, AZ: Coast Aire Publications, 1997.

Graham, Brian. *Geography and Air Transport.* Chichester, UK: John Wiley & Sons, 1995.

Holloway, Stephen. *Straight and Level: Practical Airline Economics.* Burlington, VT, and Aldershot, UK: Ashgate Publishing, 1993.

Radnoti, George. *Profit Strategies for Air Transportation.* New York, NY: McGraw Hill, 2002.

Shaw, Stephen. *Airline Marketing and Management, 4th Ed..* Burlington, VT: Ashgate Publishing, 1999.

Tretheway, Michael and Oum, Tae. *Airline Economics: Foundations for Strategy and Planning.* Vancouver, Canada: University of British Columbia, 1992.

Wells, Alexander T., and Wensveen, John G. *Air Transportation: A Management Perspective, 5th ed..* Belmont, CA: Thomson-Brooks/Cole™ , 2004.

# Chapter 14
## Aircraft Operating Strategy

**Chapter Checklist**

- Route and strategy selection

- Aircraft selection

- Aircraft type availability

- Aircraft capacity

- Aircraft range

- Fleet and crew commonality

- Aircraft commonality

- Powerplant selection

- Aircraft performance

- Aircraft efficiency

- Maintenance issues

- Hull and liability insurance

- Noise restrictions

- Noise and operational curfews

- Aircraft economic & performance evaluation

- The purchase / lease decision

**Introduction**

Deciding what type of aircraft to utilize is one of the most important decisions management must make when starting a new air carrier or expanding an existing company. All too often, management makes the wrong decision when it comes to aircraft operating strategy. The most important lesson one should know is to never select an aircraft based on its aesthetic appeal. In other words, never select an aircraft based on how attractive it looks or how fast it can fly. Never decide what type of aircraft to operate until the route network has been determined. Many airlines continue to fail because they have not selected the right aircraft for the right route. For example, it does not make sense to fly a long-haul wide-body aircraft with four engines on a route of 500 miles because the operational costs exceed any profit that could be earned utilizing a smaller more cost effective aircraft. That being said, some of the larger airlines do operate large aircraft on short-haul routes for the purposes of training or repositioning the aircraft on a long-haul flight. There is some marketing strategy involved in such an operation but that subject is better discussed in books focused on airline marketing strategy. This chapter outlines the main factors involved in deciding what type of aircraft to operate. Topics include: route and strategy selection, aircraft selection, aircraft type availability, aircraft capacity, aircraft range, fleet and crew commonality, aircraft commonality, powerplant options, aircraft performance, aircraft efficiency, hull and liability insurance, restrictions and curfews, maintenance issues, as well as aircraft performance and evaluation.

**Total Page Length: Approximately 4 to 6 Pages**

**Route and Strategy Selection**

Perhaps the most significant decision to be taken within an airline – one that will affect all others – is the type of route network and network strategy that the airline wishes to obtain. It is also fairly safe to assume that, once a strategy is decided upon and implemented in the real world, it becomes hard to disassociate the airline from the strategy with which it follows at the beginning of its life.

The type of route network selected – short-haul, medium-haul, long-haul, or a combination thereof – and the specific routes in which the airline will compete, will have a significant effect on operations, as will the decision as to what type of carrier the airline will be (For example, low-cost or premium).

It is also necessary to identify at this time potential routes for initial service as well as routes that could be added (or conversely, dropped) should market factors, such as passenger demand and competitor behavior, change in the future. If one thing above all else can be guaranteed in the airline industry, it is that things in the future will likely be significantly different from the situation at the present time. The pace of civil aviation change is so rapid that oftentimes, it is unrecognizable from one decade to the next.

In addition, another crucial decision is the composition of the fare and rate structures, and their implications of service pricing. For example, some airlines choose to charge flat fares for all seats, without the use of even basic yield management techniques that ignore the consumer surplus of consumers (where consumers are wiling to pay more than they eventually do for the service). Conversely, most airlines choose to go with, at the very least, basic yield management systems that categorize consumers according to either time or price sensitivity, and charge different fares accordingly.

## Aircraft Selection

A major step to be taken in the aircraft selection process is the identification of the needs of the routes on which the aircraft is to be flown. Issues such as passenger and cargo capacity, utilization rates and range requirements are all factors in such a selection. In addition to aircraft applicability and selection, engine (powerplant) selection may also play a role in airframe selection.

## Aircraft Type Availability

Aircraft that are generally bigger, offer greater range, and at a lower seat-mile cost than their smaller counterparts. Boeing, from smallest to largest, offers the 717, 737, 757, 767, 777 and 747 lines. Airbus, meanwhile, offers the A320, A310/300, A330/340 and A380 series of aircraft. With the exception of the Boeing 747 and A300, all of the above aircraft are offered in more than one length, trading range for capacity where appropriate. In most cases, these stretches or shrinks overlap their passenger capacity with other aircraft in the manufacturer's family, but will offer different maximum take-off weight (MTOW) and range combinations. For example, the 757-300 and 767-300 offer similar capacities, with the 757-300 designed more for medium-haul routes, while the 767-300 is designed for longer routes, up to 6,000nm in the case of the 767-300ER. The 747 and A380, 777 and A330/340, 767 and A300/310, and 737/757 and A320-series, compete with each other, with the 717 competing largely with the new A318 and A319.

Each of these aircraft offers advantages over another, but at the tradeoff of perhaps some other characteristic. For example, the Boeing 777 is offered with folding wings in order to fit in certain smaller gates. The reason for this is because the 777 was designed to essentially replace and upgrade L-1011, DC-10 and MD-11 aircraft, in addition to the competitor's aircraft. These aircraft had smaller wingspans (the distance from wingtip to wingtip) than the 777, but gates were built with the design of the DC-10 used as a basic guide. Thus, the 777 is offered with folding wings in order to ensure that, should customers require this option, they can directly replace the former aircraft, without fear of gate space issues. To date, however, no operational airline has ordered the 777 with the option of folding wings.

In the past, the majority of long-haul routes were flown only between major hubs with large aircraft such as the Boeing 747, allowing connections to smaller, secondary cities through connections with either the same carrier or other carriers, but on smaller aircraft. For example, in the past, a passenger flying from Atlanta, GA to Manchester, UK might have flown first to New York, then to London, and then to their final destination. Today, the trend has been partially reversed. With the proliferation of secondary hubs and smaller aircraft, new services from secondary hubs – or point-to-point hub-bypass services – have become commonplace. Today, the passenger wishing to fly from Atlanta to Manchester can do so directly. In addition, services are now appearing, partially through the development of regional jets such as the Bombardier CRJ-200 and Embraer ERJ-170, to avoid hubs altogether.

On any given collection of routes, it may be necessary to analyze the use of specific aircraft types by the competition. This is because passengers gravitate, as a rule, toward newer, more spacious jet powered aircraft with more passenger amenities. For example, a scheduled airline flying the Atlantic will likely find itself falling behind in passenger loads if it fails to address the competition's use of more generous legroom, personalized in-flight entertainment (IFE) and bigger, more luxurious seats with more features in premium classes. Older aircraft can generally not support modern IFE systems without major, expensive modification. As a result, many older aircraft are being replaced on premium and prestigious routes as other airlines introduce newer aircraft.

## Aircraft Capacity

The first, and perhaps most immediate, issue to be addressed in aircraft selection is that of capacity. For small routes with, for example, 200 passengers, a Boeing 747 (350-500 passengers) would not be the appropriate airliner. Meanwhile, for a route estimated to have demand for 400 passengers per flight, a Boeing 767 (170-250 passengers) may also not be appropriate. As such, the result is that route demand should be accurately forecasted in order to select the optimal size of the aircraft. In addition, the capacity of the aircraft will vary according to the type of services provided. If the operation, for example, is that of a scheduled charter carrier, with high-density one class seating optimized for routes of short to medium duration, then it follows that the number of seats on the aircraft will be higher than the number of seats in a similar aircraft, flying for an airline offering long-haul, three-class service, with first, business and economy class sections. This relates directly back to the business strategy and market differentiation techniques outlined previously.

In addition to the passenger carrying capacity of any aircraft, it is also important to note and take into account the cargo carrying capacity. A good portion of the revenue generated on scheduled flights often comes from the cargo being transported in the lower deck cargo holds. Routes have been flown in the past where the cargo provided the profits for the airlines, and the passengers provided only additional revenue on top of that. While the reality is no longer as appealing, it is true that the cargo carried on board a flight may make the difference between making a loss and breaking even, or even making a profit.

## Aircraft Range

Another major factor in aircraft selection is the range of any specific type of aircraft. Depending on the type of services offered and the route structure of the firm, the company will fly either short-, medium- or long-haul services, or a combination thereof. As such, it is necessary to have aircraft that are designed for the duration and frequency of services that the company intends to provide. Most immediately, most long-haul aircraft – those with ranges exceeding 4,000 nautical miles or thereabouts – are designed for operations which do not require many repeated cycles (a takeoff and landing) in one day. That is to say that a typical Boeing 747-400, for example (deliberately taking note and excluding those 747s designed solely for short-haul services in the Japanese market – most notably, the 747-400D) is designed for many hours of flying per day, but not numerous takeoffs and landings. As a general rule, the longer the range with which the aircraft is designed, the less efficient the aircraft will be at short-haul routes.

Conversely, smaller aircraft – such as the Boeing 737 and Airbus A320 – are designed for flights of lesser distance, but increased frequencies. The essential difference here lies in the construction of vital physical parts such as undercarriage and pressurization system – which are built for more repeated use than those aircraft designed to fly longer, less frequent routes.

There are aircraft, however, of larger size, that are designed for short- or long-haul operations. Good examples of this are the Boeing 767 and the Airbus A300. These aircraft have the range and capacity capabilities of a large aircraft (in their –ER models, at least), but also featured the strengthened pressure bulkheads, landing gear and other items of shorter-haul aircraft.

The next implication of route type also affects aircraft selection. As a general rule with a few notable exceptions, long-haul aircraft tend to be larger than their short-haul cousins. While aircraft such as the Boeing 767 and Airbus A330 fly not only international, but also domestic, routes, aircraft larger and smaller have diverging purposes and goals. As a result, short-haul aircraft tend to be narrow-bodied aircraft (of Boeing 757 size and lower) while wide-body (twin-aisle) aircraft, such as the Boeing 767 and Airbus A310, are preferred by passengers and the airlines for long-haul services. Passengers prefer them because they are perceived as being more spacious for long stretches than their narrow-body cousins, and airlines prefer wide-body aircraft for their notably increased cargo carrying capacity. With the exception of the Airbus A320 line, wide-body aircraft accommodate cargo and baggage containerization (using pallets and standard Universal Load Device (ULD) containers) where narrow-body aircraft do not.

**Fleet and Crew Commonality**

Fleet Commonality and two-pilot crews are also advantages with modern aircraft, over older aircraft. Fleet commonality between aircraft such as the Boeing 757 and 767, or Airbus A319, A320 and A321, enable airlines to save money through the use of consolidated repair, spare-parts, and crew pools, hence saving money due to the lack of repetition and duplication. The 757 and 767 share almost 60% of their parts with each other, despite their dissimilar appearances, and the A319/320/321 share much more than that.

Most aircraft built today (and all commercial aircraft from the major Western equipment manufacturers) offer two-crew cockpits, reducing both the number of cockpit crew required from the previous three, and crew workload, through the use of aircraft system automation and electronic flight information systems (EFIS). This is commonly referred to as, "glass cockpit" technology. Further, modern aircraft also sometimes offer cross-crew commonality between types. For example, with only a few days of conversion training, a pilot can be simultaneously certified on the Airbus A320-series and Airbus A330/340 series, or on the Boeing 757 and 767. While some airlines in some parts of the world have been able to take advantage of this immediately, the approach in the United States has been one of caution, where contracts between airline and employees have limited the scope of any crew from simultaneously flying two types of aircraft within an airline's fleet, largely due to seniority and pay-scale issues.

## Aircraft Commonality

For airlines with aircraft in existing service, aircraft commonality and cross-crew qualification ability are significant factors in adding new types of aircraft, either derivates or completely new airframes. For airlines that intend to expand in the future, but wish to start small, it is an obvious advantage to begin with an aircraft type that offers crew and part commonality with aircraft that are either smaller or larger, depending on planned expansion requirements, to minimize the outlay and investment required at that time when it becomes necessary to grow and diversify into new equipment types. Commonality among types reduces the total cost associated with the introduction of a new type of aircraft into service. Maintenance and crew training costs are major costs that must be factored into the equation when adding new types of aircraft, and as such, it is economically advantageous to minimize these costs by avoiding unnecessary duplication.

Powerplant similarity and commonality is also an operational advantage. For example, companies that operate the Boeing 757 with the Rolls Royce RB.211-535 engine, can take advantage of the significant commonality that this powerplant offers with similar RB.211 engines found on Boeing 767, 747 and Airbus A330 aircraft. This should also be a major factor, as previously discussed, in airframe/powerplant combination and selection.

## Powerplant Selection

Powerplant selection is also a major decision in aircraft acquisition. Aircraft characteristics can vary significantly (by 1 or 2 percent) between powerplant performance, and each different available powerplant will have various different characteristics and requirements.

For some aircraft, there is only one type of engine available – for example, the Boeing 737-800, available only with a version of the GE/SNECMA CFM-56. Meanwhile, the Airbus A320 is available with either a similar CFM-56 or the International Aero Engines (IAE) 2500. The Boeing 767, meanwhile, is available with engines from Pratt & Whitney, General Electric and Rolls Royce. Some engines may be physically heavier than others, but offer more efficient fuel consumption. Some may be cheaper to purchase, but offer less efficiency in the long run.

Maintenance issues with engines can be defined as either number of shop visits per 1,000 flight hours or cycle limits between maintenance shop visits. The lower the number of shop visits per hour, the longer the time between shop visits, as a function of engine flight hours (EFRs).

For Extended Twin-Engine Operations (ETOPS) purposes, different aircraft/powerplant combinations have different certificates and levels of performance. An ETOPS certificate is awarded to the combination of engine and airplane, rather than the aircraft type itself. This is an additional consideration for aircraft/powerplant selection processes.

## Aircraft Performance

Once range and capacity requirements are established, it is then necessary to look at and analyze the performance characteristics of aircraft. As a general rule, modern aircraft are more efficient than their older counterparts. While older aircraft may be available for less initial expense (such as the Boeing 737-200 or Douglas DC-9) than more modern, fuel-efficient aircraft (such as the Boeing 737NG or Airbus A320), the operating costs and penalties associated with operating such aircraft may negate their initially lower acquisition costs.

It is important to note at this point, however, that it is crucial to be familiar with all different combinations of aircraft and powerplant and, more fundamentally, all the available types of aircraft available for each type of route/payload requirement.

## Aircraft Efficiency

Modern aircraft can often do the job of older, less efficient aircraft with less fuel, less crew and increased passenger and cargo capacity. For example, the Boeing 777 can fly the same route as the largest Lockheed L-1011 TriStar (the L-1011-250), but with only two engines as compared to three, a two-crew cockpit instead of the original three, and with less fuel consumption than its predecessor. In addition, the engines of the Boeing 777 are not only more powerful, but also quieter, than those engines used on the L-1011 and other older aircraft.

Newer aircraft are also more fuel-efficient. This can make new aircraft – in preference to older, less efficient aircraft – more appealing in times where fuel costs are relatively high, or where they are set to rise in the near future. In addition, other factors that affect the preference whether to acquire new or used aircraft can be the resale value of new aircraft at the end of their expected useful life with the carrier and the availability of, and demand for, used aircraft in the marketplace.

## Maintenance Issues

Considerations to take into account in aircraft operations must also include those of maintenance. As a general rule, a new aircraft can operate for roughly 7 to 10 years without the need for major structural repairs.

With that in mind, it is also important to note that, as a percentage of total operating expenses associated with aircraft operations, maintenance costs have been gradually decreasing since the advent of the jet age. Where once, in 1960, maintenance costs accounted for roughly 19 percent of total costs, the number was only around 11 percent in 1995.

In the United States, maintenance issues fall under the jurisdiction of the Federal Aviation Administration (FAA) and the Flight Standard Service (FSS). Aircraft maintenance issues are tightly regulated, and there are specific guidelines by which airlines must follow. The maximum limits between these checks are referred to as Hard Time Limits.

Service Checks are basic maintenance checks including a general inspection of the interior and exterior of each aircraft, which is performed before each and every flight, by either a member of the flight crew or a member of the ground crew, while the Time Between Overhaul (TBO) is the time between major checks, such as C and D checks.

A-checks, performed at a maximum of every 155 hours and usually done overnight, consist of a more detailed visual inspection and the repair of any necessary items that have arisen.

B-checks, performed every 750 flight hours, also occur overnight and include the removal of service panels and the inspection of cowlings, oil filters, and a more detailed examination of the airframe.

C-checks, performed every 3,000 flight hours, take roughly 5 days and consist of detailed inspections, removing any outer surface to investigate further, and include the calibration of any and all flight surfaces, and engine and internal mechanism tests and rechecks.

D-checks are significantly more complicated, take significantly more time (up to one month), and happen more infrequently (usually between 15,000 and 30,000 flight hours). They include a major reconditioning of the entire aircraft, where it is essentially disassembled and reconstructed, piece by piece, to establish any corrosion or other damage that may have occurred.

Some items within maintenance checks can be categorized as structurally significant (SS) items, which mean they must be investigated and repaired at the nearest possible opportunities.

A major operational decision concerning maintenance is the decision as to whether to perform all maintenance in-house, outsource all maintenance to either another airline or company and, in addition, where that maintenance should be performed. In addition, if maintenance facilities are to be located at the home base, it should also be determined what exact services will be performed at that base. It is unlikely for a start-up airline that all maintenance up to a D-check will be performed in-house at the maintenance base. Few airlines offer D-check service checks, and the majority of smaller airlines outsource this service to the major airlines, such as those in Atlanta, London, and San Francisco.

## Hull and Liability Insurance

Hull and Liability Insurance, as in many industries, is vital and legally required for aircraft operators (airlines) in order to protect the assets of both the operator and those who travel in the aircraft. Hull insurance is that insurance which covers against the cost of repair or replacement of the aircraft itself, and related systems, should they be damaged in any way in an accident. Conversely, liability insurance covers the airline operators in the event that a passenger or passengers are injured, killed or otherwise harmed during an accident while onboard that aircraft. Aviation policies generally divide this liability coverage into general liability – that which covers all aspects but passengers – and passenger liability, covering only passengers. Death, injury, damage, loss, and terrorism are important subjects to explore in greater detail once the airline business plan begins to be developed.

**Noise Restrictions**

Since the late 1990s, most western nations (including the entire European Union and United States) have imposed stringent noise regulations on civilian jet aircraft operating within their airspace. These new restrictions prohibited any aircraft that failed to meet Stage III (also called Chapter III in certain country) noise regulations, from operating.  Stage I aircraft include all those "first generation" jetliners, including the majority of Boeing 707s, the original powerplant Douglas DC-8, Vickers VC-10, deHavilland Comet (and its descendents), the Sud Aviation Caravelle and the BAC/Aerospatiale Concorde.

Beginning the early 1960s, aircraft became slightly quieter, and these are referred to as Stage (or Chapter) II aircraft. These include the McDonnell Douglas DC-9, original Boeing 727 and 737 (-100 and -200), and the BAC 1-11. The engines that powered these aircraft were slightly higher in bypass ratio (the amount of cool, slow moving air around the hot, fast-moving engine core air) than their first generation counterparts.

With the advent of wide-body jets and the subsequent high-bypass turbofan engines that were developed in tandem with them, the noise level produced by these engines fell further still. These aircraft – including all western wide-body jets and the majority of aircraft produced after 1980, with the exception of the Boeing 727 – are classified as Stage III aircraft. Modern aircraft, such as the Boeing 747-400, 777 and Airbus A330/A340 are quieter still, and fall into the Stage (Chapter) IV category.

The introduction of these stringent noise restrictions in the majority of the western world over the past ten years, particularly in relation to a minimum Stage III noise restriction, has resulted in the mass market exodus of many older aircraft such as Boeing 707 and Douglas DC-8. Earlier Boeing 727 and Douglas DC-9 aircraft were also affected by this. While some DC-8 and 727 aircraft were re-engined by UPS and others using GE/SNECMA CFM-56 and Rolls Royce Tay engines respectively (effectively bringing them up to Stage IV standards), another market also arose, in hush-kitting.

Hush-kitting is the process of adding an additional part onto the back of an engine of a Stage II aircraft, such as a 727-200 or DC-9-30 – in a similar fashion to the silencer on a pistol – to reduce the level of noise coming out of the back of the aircraft.  Essentially, this will slightly reduce performance, but it is noticeably less expensive than replacing either the engines or the whole aircraft. It brings former Stage II aircraft into line with Stage III noise requirements, but neglects to tackle other issues, such as the emissions from the older engines.

Despite these operational noise restrictions, one of the few exceptions made was for the Stage I-compliant BAC/Aerospatiale Concorde, and then only at New York JFK and a few diversionary airports, such as Bangor, ME and Gander. Other aircraft must meet the noise restrictions.

It is important to note, however, that some European countries prohibit the use of hush-kitted Stage II aircraft, despite their Stage III noise compliance, due to the reduced engine and fuel efficiency associated with the installation of these hush-kits. These governments argue that hush-kitted aircraft are still as polluting as the original airframe, and only produce less noise. Due to this, therefore, they ban them from operating within their borders.

## Noise and Operational Curfews

A result of noise compliance in many cities across the western world has been the imposition of noise curfews. The effect of this is that certain aircraft are banned from operating from certain airports, and others are banned from operating at certain times of day. Many airports in the United States and around the world have evening operating restrictions, dictating alternating runways at certain intervals or, worse, prohibiting aircraft movements after a certain time in the evening (usually 2300 or midnight) and before a certain time in the morning (usually between 0500 and 0600). In London, for example, airlines can incur significant fines for arriving too early from across the Atlantic at either Heathrow or Gatwick, and often circle high above the city for up to forty minutes, in order to not land short of the curfew.

These restrictions affect passenger operations perhaps less than dedicated cargo services. While most people sleep, air cargo companies are busy at work sorting, handling and flying cargo across the globe. The result is that cargo companies often select airports where curfews are either short (i.e., 0100-0400) or non-existent. Airports where this occurs are often outside of major metropolitan areas and, as such, also often provide large areas of space in which cargo operators may place sorting and dispatch facilities, and aircraft stands. An example of this is Kansai International Airport, located offshore outside the Japanese city of Osaka. The result of this is that operations can take place 24 hours per day, 365 days per year.

Other operating restrictions can come from slot restrictions that limit airport access at certain airports, bilateral agreements with foreign countries, and crew and scheduling limitations.

Slot restrictions are not always common in the U.S., but are very common in other countries. Only four major airports in the U.S. have slot restrictions (New York JFK, New York LaGuardia, Chicago O'Hare, and Washington National). However, airports in other countries – such as the entire European Union – are moving toward an entirely slot-based system, whereby airport access can be awarded, traded and otherwise revoked to or with any specific airline.

Bilateral agreements are currently in a constant state of flux. A modern trend, Open Skies (the right of any carrier of either of the two concerned countries to fly between any two cities in the two countries), has dominated recent bilateral agreements.  However, with the European Union now negotiating bilateral aviation agreements as one unit (for example, the U.S./E.U. bilateral talks) the future of these agreements has yet to be determined. It is likely that, at the very least, the agreement between the two countries will yield Open Skies, allowing any U.S. carrier to fly from anywhere in the U.S., to any point in the E.U. – or vice versa.

Other operational considerations include pilot and crew workload limitations. Most western nations – and all those that fall under either Joint Aviation Authority (JAA) or (Federal Aviation Administration) FAA jurisdiction – have limits on both maximum daily, monthly, and yearly flying time for commercial pilots, and also maximum duty time, including any airport flight-planning time. These regulations also specify the minimum rest time allowed, by law, for pilots operating from bases in the specific country.

## ETOPS

However, the issue of two-engine operations for extended flights also raises an issue which previously, with three- and four-engine aircraft, went unaddressed. The Boeing 757, 767, 777 and Airbus A310 and A330 all qualify for a rating known as Extended Twin-engine Operations (ETOPS), providing long-haul operations, usually over the oceans, while never operating more than a certain distance (measured in flying time) from an alternate landing field, should one engine fail.  It is a function of both airframe and powerplant combination. That is to say that one aircraft-powerplant combination may have an ETOPS certificate, while another powerplant on the same aircraft may not have the certification.

For example, when it entered service, the Boeing 767 qualified only for 60-minute ETOPS certification, meaning that it could only fly up to 60 minutes from a potential diversionary airport, should one engine fail. However, with time and experience, it became apparent that the aircraft's engine reliability rate was high enough to permit operations up to 180 minutes from an alternate landing field. Conversely, the Boeing 777 was introduced into service (after extensive pre-service testing) with 180-minute ETOPS certification. Once established, Boeing and the engine manufacturers (since ETOPS certification is a function of both aircraft and powerplant combination) sought to increase Boeing 777 ETOPS certification from 180-minutes to 207-minutes, closing up the last gap in the Pacific Ocean, allowing non-stop B777 flights from the U.S. West Coast to Australia and New Zealand.

## Aircraft Economic and Performance Evaluation

As previously discussed, the quantity of both passengers and cargo that an aircraft can carry – and the range over which it can carry them – has a major impact on the selection of aircraft types available to an airline, for a given route network. This is measured, primarily, through the weight of payload that can be carried within the aircraft. The equation to determine the payload requirements for passengers can be written as follows:

(Number of seats on airplane * Passenger weight) +
(Number of passengers * weight of baggage per passenger)

In almost most all situations, it is inappropriate to ask for passenger weights at check-in. Therefore, it is often the case that an average weight per passenger is assigned (roughly around 180lbs~200lbs / 80-90kg), along with the average weight of baggage (usually estimated around 40lbs~50lbs / 18-22kg).

In addition, it is also necessary to determine the average weight of cargo carried on the route, and add it to the total payload requirements of the passenger load. This can be established by determining the average amount of cargo, and the tare weight of the containers and pallets upon which it might sit. In addition, the physical volumetric space required for both baggage and passengers must also be considered. Some aircraft have excellent weight-carrying abilities, but sometimes suffer from volumetric limitations – or vice versa.

To determine the economic and operational implications of a certain type of aircraft, it is also necessary to establish the exact payload (both weight and volumetric) specifications of the aircraft. The information to be determined will include the following.

The range/payload tradeoff must be considered. For every aircraft type, the greater the payload, the reduced range that the aircraft can fly. Conversely, for a route of any given range, it is necessary to adjust payload accordingly, depending on atmospheric conditions, weather considerations and prevailing winds. For example, a 3,300 mile flight from Denver to Quito, Ecuador (both very high altitude cities) in summer, when the air is hot and dry, would require significantly more fuel for take-off and landing than a flight of similar length from London to New York in winter, where the air is cold and dry. As a result, the payload on the Denver-Quito flight would have to be significantly reduced or, conversely, the fuel load significantly increased.

A primary cost of operation with any aircraft is its Direct Operating Cost (DOC). This can be measured in total trip cost, direct operating cost per mile (measured in $), or by direct seat-mile cost, measured by taking the total direct operating cost per mile and dividing it by the number of seats on the airplane. Ideally, every seat on the aircraft should be filled with a passenger, but this rarely happens. Thus, the Cost per Available Seat Mile (CASM) is a minimum cost, while a reduced load factor will result in higher per-seat costs for any specific flight.

From the determination of the aircraft's CASM and the fares to be charged on the route (as determined by revenue management or matching up to equivalent competitor fares), it is then possible to determine the Break-even Load Factor for the route. This can be measured by the total cost per trip, divided by the total revenue generated from the trip.

For cargo, the cost associated with transportation is measured by Cost per Available Ton Mile (CATM). The revenue generated can be measured in Revenue per Available Ton Mile (RATM) or revenue per flight. The result is that cargo, while often not factored into the CASM equation, can provide additional revenue for any flight and, in some circumstances, make the difference between making a loss, and yielding a profit on the flight. It would likely be advantageous to include cargo revenues where applicable in any projected revenue from a flight (with the research associated with the size of the cargo market on that route) to ensure that revenue projections are as accurate as possible.

It is also necessary, before any aircraft decisions are undertaken, to establish which routes provide an opportunity for economic profit and have sufficient passenger demand to warrant service. It must be noted at this point, however, that it is crucial that no route is served merely for sentimental reasons. As with any aircraft purchase, any route that cannot sustain its existence through economic justification, should not be considered for service. As a side note, it would also be hard to justify any such route to potential investors, when they see the lack of economic profit (and hence, return on investment) from the route. In addition, it is also crucial that no aircraft type is selected solely, or primarily, on the basis of its aesthetic appeal. The best looking aircraft may not be the best aircraft for the airline/route/payload combination which the company wishes to undertake.

By extension, it is necessary to establish or acquire accurate and timely marketing information and forecasts for any growth, stagnation or reduction in route size, passenger numbers, or change of passenger demographics in the short- to medium-term future. This information is crucial not only to determine the ultimate success of the route for internal purposes, but also for the determination as to viability of the route (and airline) for potential investors.

## The Purchase / Lease Decision

When acquiring new aircraft, airlines face the choice of either purchasing or leasing those aircraft. Before the 1980s, the majority of airlines purchased their aircraft. However, the trend is now of leasing and purchasing, and over half of all new aircraft are leased by the airlines, rather than purchased. Major airlines now lease significant numbers of aircraft within their fleets. Many have purchased aircraft, only to sell them to leasing companies and immediately lease them back. This is appropriately termed, "sale and lease back." Major lessors (those that own the aircraft and lease them to the airlines) include GPA Group (formerly Guinness-Peat Aviation) of Shannon, Ireland, and International Lease Finance Corporation (ILFC), based in Beverly Hills, CA.

There are essentially two types of leases. **Operating leases** are non-cancelable leases where the lessor retains full title to the aircraft and assumes all post-lease risks as to the value of the aircraft. The advantage of this is that the operator of the aircraft (the lessee), has no obligation beyond the end of the lease to the aircraft, and there is no significant initial cash outflow associated with the lease.

The second basic type of lease is a **Financial (Capital) lease.** With these, title remains with the lessor until all payments have been completed. Title then changes from the lessor to the lessee, and a final lump sum is paid at that time to the lessor. Unless the lessee defaults, therefore, there is no danger to the lessor because they do not assume the risk associated with the value of the aircraft following the end of the period of the lease.

In both cases, responsibility for the upkeep and operation of the aircraft, including any maintenance and insurance, usually fall to the lessee. The exception to this rule is the short-term Aircraft, Crew, Maintenance and Insurance (ACMI) lease, where the lessee assumes only the direct operating costs of the aircraft and the lease payments. These arrangements tend to be of short-term duration and are often done by one airline for another. Ad-hoc cargo services by major airlines sometimes operate on this basis.

The success of aircraft leasing has now made the leasing companies, including the GPA Group and ILFC, some of the aircraft manufacturer's largest customers. For small carriers that lack the capital capability to purchase large aircraft, leasing offers them a method by which to acquire and operate them.

In today's global market of plentiful amounts of aircraft leases, it is often true that many aircraft are registered outside of the country in which they intend to operate. For example, many US-registered aircraft are permitted, through clauses in bilateral agreements, to operate services, under lease, for foreign carriers, within the foreign carrier's home territory. For example, Air Atlanta Icelandic often leases aircraft under the Icelandic registry (TF-) to airlines in the United Kingdom, Europe, and beyond.

In addition, freight aircraft in the U.S. frequently operate cargo services on behalf of foreign carriers. However, at this time, agreements between certain countries mean that the US does not permit foreign-registered aircraft to operate within the United States. Indeed, this continues to be an issue in the ongoing talks concerning aviation agreements between various countries.

## Additional Miscellaneous Factors

As an addition, any additional factors not discussed above that are or may become relevant to either route or aircraft viability in the future, or the general operations of the aircraft, should be discussed in this section of the business plan.

## Summary

Selecting an aircraft is an extremely time consuming process but a worthwhile exercise. Historically, many airlines have operated the wrong aircraft on the wrong route resulting in high operational costs and inadequate seating to match fluctuations in demand. This chapter discussed the basics behind selecting the right aircraft for the right route and introduced the reader to the different factors that must be considered when flying short-haul, medium-haul, or long-haul routes. Major topics covered included route and strategy selection, aircraft selection, aircraft type availability, aircraft capacity, aircraft range, fleet and crew commonality, aircraft commonality, powerplant options, aircraft performance, aircraft efficiency, hull and liability insurance, restrictions and curfews, maintenance issues, as well as aircraft performance and evaluation.

## Key Terms

Route network requirements

Fare and rate structure

Aircraft characteristics

Fleet/crew commonality

Maintenance issues

Extended Twin-range Operations (ETOPS)

Purchase/lease decision

Noise and operational curfews

## Review Questions

1.  What is involved in the aircraft selection process?

2.  How do ETOPS restrictions dictate the type of aircraft suitable for airline purposes?

3.  What are the implications associated with purchasing or leasing an aircraft?

4.  How does powerplant selection, where applicable, affect aircraft performance characteristics?

5.  What are the different types of maintenance checks required by law to keep aircraft flying?

## Web Sites

http://www.boeing.com/commercial

http://www.airbus.com

http://www.ilfc.com

http://www.geae.com

http://www.rolls-royce.com/civil/

http://www.pratt-whitney.com


**Recommended Reading**

Abeyratne, Ruwantissa. *Aviation Trends in the New Millennium.* Burlington, VT: Ashgate Publishing, 2004.

Radnoti, George. *Profit Strategies for Air Transportation.* New York, NY: McGraw Hill, 2002.

Shaw, Stephen. *Airline Marketing and Management, 4th Ed..* Burlington, VT: Ashgate Publishing, 1999.

Tretheway, Michael and Oum, Tae. *Airline Economics: Foundations for Strategy and Planning.* Vancouver, Canada: University of British Columbia, 1992.

Wells, Alexander T., and Wensveen, John G. *Air Transportation: A Management Perspective, 5th ed..* Belmont, CA: Thomson-Brooks/Cole™ , 2004.

# Chapter 15

## Competition and Competitive Response

**Chapter Checklist**

- Main competitor identification

- Identification of other competitors

- Market strategy

**Introduction**

Regardless of the type of industry, virtually all businesses face competition. This is a reality one must accept and appropriately cope with. It would be a fatal mistake to disregard it, especially during the planning phase of a new business meaning that this subject must be appropriately described and evaluated in the business plan.

Information on competitors contained in the business plan must cover both the current forms of competition, current trends, as well as potential competitors affecting the new airline. In order to make that information easily understandable for the reader, it is highly advisable to break information down into the competitors′ strengths and weaknesses. Depending on the selected type of airline and the region it will operate in, the airline business plan should produce analyses for three to five competitors with a main focus on the primary one.

The next logical step rests in a description of a market strategy the new airline will use in order to stay ahead of its competition and to maintain that position. This point is extremely important, especially in a new era of aviation where the global airline industry is going through a lot of significant changes affecting every sector of air transportation.

This chapter summarizes the main elements of dealing with competition and competitive responses and encourages the reader to make extremely thorough and deep analyses of current and future competition through the following blocks: main competitor identification, identification of other competitors, and market strategy.

**Total Page Length: Approximately 2 to 4 Pages**

**Main Competitor Identification**

The analysis of any new airline's competitors starts with identification of its main competitor. In the past, there were some startup airlines that skipped this important step and the price they paid was high – bankruptcy. They were forced to leave the market because they did not have sufficient knowledge on their primary competitors, and as a result of that, they logically did not take all the necessary measures that might have kept them in the business.

The primary reasons of their failure were associated with their lower operational efficiency, lower quality planning capabilities and schedules, wrong decisions on obtained equipment, insufficient market analyses, incorrect market development projections, and most of all, poor risk analyses.

In order to make sure a new airline gradually becomes profitable, none of the aforementioned areas should be overlooked. As much information as possible concerning the primary competitor should be obtained.

The primary competitor's assessment should include the following items: summary of product lines, strengths and weaknesses, sales and estimates, primary competitor's targets, and its impact on the new airline. Once this information is known, the airline business plan should be able to tell the reader why the new airline will be more successful versus the competition. Adding uniqueness to the airline helps address this issue.

All the mentioned types of information presented are publicly available. The most convenient sources of information include airline databases, official airline industry reports and reviews, including government sources, and information published by the primary competitor. One should not exclude the possibility of contacting the primary competitor directly and requesting the necessary information.

When all the needed information becomes available, one has a nice foundation for further evaluations and analyses, but before they commence, the other, less critical competitors, should not be overlooked. Why? As it has been already mentioned, the current airline industry is extremely dynamic and that is why things change. A small company that is small today may become big tomorrow and vice versa.

**Identification of Other Competitors**

Identification of other competitors rests in their detailed assessment which is basically similar to the main competitor's assessment. The business plan must provide its reader with a sufficient amount of information on all the major competitors and their services the customer might buy instead of the services provided by the new airline.

The reader must learn the number of other competitors, how large they are, including their relationships towards the target market, expected market share, advertising methods and promotion used by the other competitors, and their possible response to the new airline's market entry.

Besides direct competition, there is also the possibility of indirect competition represented by other businesses or airlines that might become competitors after one enters the market. Basically, one needs their cooperation or tolerance, and that is an issue requiring attention.

The business plan reader will also be interested in features that make the new airline different from its competitors, including all the associated advantages and benefits.

Finally, a great deal of attention must be devoted to air ticket prices. The airline business plan must provide answers to a number of questions. For example, will prices be competitive, under what conditions, and why?

## Market Strategy

Once a sufficient amount of information has been collected concerning the main competitor and other competitors, analyses of data may start. Upon completion a market strategy becomes available. The business plan reader will be interested in knowing how the new airline plans to successfully handle its current and future competitors operating in its market segment, how much it will cost, and what the possible alternative courses of action will be.

In order to provide a complete set of answers to the above, the business plan should explain the new airline's strategies concerning possibilities of increased differentiation, its scope of flexibility in terms of reacting to changing market conditions, including a potential market withdrawal plan.

## Summary

The process of analyzing competition and producing a competitive response is not an easy task. This is extremely important research that must be done in the best possible way, since it will form a foundation for further planning steps and very critical financial decisions. Thorough and deep analyses leading towards a professional description of competitors that the new airline will have to deal with will not only satisfy the investor's questions and doubts, but will also contribute to the future success of the airline. As mentioned in a previous chapter, airlines often make too many assumptions when it comes to competitive issues. Many airlines have failed in the past and continue to do so today because they often assume there is no competition. Management should never assume there is no competition and if it does exist, never assume that one airline is better than another.

Since the terrorist attacks of September 11th, 2001, SARS, and the Iraqi crisis of 2003, the global airline industry has been exposed to competitive issues that were never addressed previously. New airlines entering the market must take into account all the existing and potential factors that impact the competitive nature of the business. Addressing such issues within the airline business plan will provide a stronger foundation leading to success versus airlines of the past that were able to commence operations with weak business and strategic plans. Investing time and the proper amount of financial resources into research associated with competition is often neglected by new airlines entering the market. This chapter informed the reader on how important research is when identifying current and future competition followed by a brief discussion on how to react to such competition.

## Key Terms

Main competitor analysis

Other competitors

Future competitors

Product differentiation

Market strategy

Alternative courses of action

Market withdrawal plan

## Review Questions

1.      Why is it necessary to deal with competition?

2.      How should be competitors described in a business plan?

3.      What are the steps in analyzing competitors?

4.      What is marketing strategy and what is its purpose?

## Web Sites

http://www.mondaymemo.net/030428feature.htm

http://strategis.ic.gc.ca/epic/internet/incb-
bc.nsf/vwGeneratedInterE/ct02181e.html

http://www.econ.umn.edu/~gautam/pdf_papers/airline_competition.pdf

http://www4.nas.edu/ocga/testimon.nsf/0/ff18a979e15724dc8525683a005a08
94?OpenDocument

http://ostpxweb.ost.dot.gov/aviation/domestic-competition/reports.htm

## Recommended Reading

Hammer, Michael. *The Agenda: What Every Business Must Do to Dominate the Decade, 1st edition.* New York, NY: Crown Business, 2001.

Hill, Charles W. L. *International Business: Competing in the Global Marketplace, 2nd edition.* Chicago, IL: Irvin, 1997.

Lawton, Thomas C. *Cleared for Take-off: Structure and Strategy in the Low Fare Airline Business.* Aldershot, UK: Ashgate, 2002.

Rosenbaum, David I. *Market Dominance: How Firms Gain, Hold, or Lose It and the Impact on Economic Performance.* Westport, CN: Praeger, 1998.

# Chapter 16

## Management and Support Team

**Chapter Checklist**

- Management team

- Organizational structure

- Corporate culture and the new workforce

- Changing nature of human resources

**Introduction**

Every new and existing organization has goals, whether they be profits, market share, growth, quality of products or services, community image, or any combination of these. Management is the process of achieving an organization's goals through the coordinated performance of five specific functions: planning, organizing, staffing, directing, and controlling. Each member of the management team must be able to perform when it comes to the five functions of management. For an airline to be successful in a constantly changing environment, the right person must be hired to do the right job, including all persons at the management level. It is important to note that investors invest in people and not in paper. The most important things to an investor are: real sales, size of market, and quality of the management team. This chapter introduces the reader to the management team, organizational structure, corporate culture, and the new workforce known as *Generation Y*, and the changing nature of human resources.

**Total Page Length: Approximately 5 Pages**

**Management Team**

This element of the airline business plan is crucial for successful operations because it helps sell the business plan to investors and other interested parties. Good management is a key to success and each participant must be able to contribute something to the business. Each position should fit with the experience and skills of the individual and each participant should be able to answer the question, "What do you offer this business venture?". As a rule of thumb, there should be at least one very experienced person on the management team. Often, such an individual is referred to as a "gray hair". To improve success of the company, the ideal person should have a proven business background, preferably as C.E.O. with a middle to large size company. The type of industry does not really matter but an airline background is a definite advantage. Describing the name, title, skills, abilities, and biography of this person along with all others selected to be a part of the management team, will help make the business concept more credible. In the United States, it is also advisable to list any person with 20 percent or more ownership as part of the management team.

Without offending the reader, there is another rule of thumb that should be considered when discussing the management team. Be cautious of the number of line pilots that make up the management team. Generally speaking, pilots do not make the most effective managers. Pilots are very educated when it comes to aircraft operations but often lack the business skills required to run a successful operation. That being said, more pilots are combining flight hours with academics and during the course of the next decade, it is expected that pilots will be more educated than in the past. In the airline industry, it is often said that there is a surplus of pilots on the market but a lack of qualified pilots in terms of flight experience combined with academic experience.

The number of people required to make up an efficient management team depends entirely on the type of operation, size of operation, and skills of the individuals. Existing airlines already have a corporate structure in place as mentioned earlier in this chapter. In most cases, the management team is too large and somewhat ineffective due to duplication of work and lack of communication between departments. New airlines starting out have the advantage of being able to establish an effective management team from the start. It is wise to have a small management team initially and grow it as the airline expands. For starters, it is recommended that the management team consist of one lead person acting as President/C.E.O. Ideally, this is the optimal position for the "gray hair" mentioned earlier. This individual is the chief executive officer of the corporation and is responsible for the proper functioning of the business often involved with the financial community, government, and members of the public. Not only should this individual have extraordinary business skills, but he/she should have good interpersonal skills as communication plays an important aspect of this position.

It is also recommended that a second lead person be a part of the management team. This person might be given the title of Executive Vice-President or Senior Vice-President. Once the airline is established and growing, it will most likely be necessary to appoint two individuals to fill each title. However, for a new airline starting out, this is not necessary unless massive rapid growth is anticipated over a short period of time. Ideally, the second lead person should have a Vice-President/General Manager title. This person should have a number of years of airline experience at the management level because he/she is responsible for the day-to-day operation of the airline.

Aside from the two positions previously mentioned, a new airline starting out might consider the following information when forming a management team. Again, keeping the initial team small is important. In the United States, for an air carrier to obtain certification, the management team must have a minimum number of positions. For FAR 121 certification, mandatory positions include: Director of Safety, Director of Operations, Chief Pilot, Director of Maintenance, and Chief Inspector. For FAR 135 certification, the mandatory positions include: Director of Operations, Chief Pilot, and Director of Maintenance. Depending on the complexity of the operation, it is possible to obtain a deviation from the required basic management positions and qualifications if requested in writing to the FAA. Such a request is normally made when the air carrier submits the formal application letter for certification. However, the air carrier must be able to show the FAA that it can perform the operation with the highest degree of safety under the direction of fewer or different categories of management personnel. The certification process is discussed in detail in *Chapter 19*. Additional information concerning the required background for each of the positions mentioned can be obtained from the FAA's Advisory Circular (AC) 120-49 entitled 'Certification of Air Carriers'.

Developing the right management team is a difficult process and in the case of many new airlines planning to commence operations, opportunities often pass by because of the length of time needed to put people in place. In order to speed up the raising of capital and move forward with the certification process, some business plans utilize the reputation of an outside party to act as the "interim" management team. There are a number of consulting companies and expert individuals in the market who will permit use of their name and talents on paper for a fee or some form of compensation. Some firms will put together an entire management team to help get a new airline off the ground. Many existing airlines have found this to be a worthwhile option but it should be noted, such an option can be expensive and somewhat risky. Be sure to obtain references for all potential members of the management team and do background searches, if necessary. The aviation industry is full of "experts" so be cautious and do not rush into any type of contract until a thorough investigation has been completed. Also, depending on who the primary investors are, they will often have a say with who should be a part of the management team. In many cases, the investor is not the best person to decide who should manage the airline. *Chapter 18* discusses investor relations in more detail.

## Organizational Structure

Most airlines, old and new, tend to operate using the classic pyramid or top-down structure consisting of top management, middle management, and operating management. There is no clear definition of each level, and meanings attached to the terms sometimes differ from one company to another. However, top management is generally considered to be the policy-making group responsible for the overall direction of the company; middle management is responsible for the execution and interpretation of policies throughout the organization; and operating management is directly responsible for the final execution of policies by employees under its supervision. The pyramid is divided into administrations each headed by an individual. For example, major units might include flight operations, marketing, or personnel. Departments are the next major breakdown within administrations; divisions within departments, and so forth. *Figure 16.1* shows a typical airline structure.

## Fig. 16.1
### Typical Airline Structure

TOP MANAGEMENT

President (Planning-Thinking Level)

Executive Vice-President (Planning-Thinking Level)

Senior Vice-Presidents (Administration Level)

MIDDLE MANAGEMENT

Vice-Presidents (Department Level)

Superintendents/Directors (Division Level)

LOWER OR OPERATING MANAGEMENT

Chiefs/Regional Managers (Section Level)

General Supervisors/Sales Managers (Group Level)

Supervisors/Sales Representatives (Unit Level)

Although this structure has been used for many years, there are different options to consider when designing the business plan. As mentioned previously in this book, the top three costs for an airline are fuel, labor, and maintenance. Increased pressure has been put on the airlines in recent years to implement cost cutting strategies and one area hit has been labor. Middle management is usually the first to be eliminated during bad times as witnessed in the mid-1980s when the United States faced a major recession. Many airlines have realized that middle level management is not always necessary to run a successful operation and new airlines often eliminate this section of the corporate structure therefore reducing costs and often improving efficiency. When middle management is cut from the picture, work ordinarily done at the department and division levels shifts upward increasing the roles and responsibilities with top management. In other cases, more authority is delegated to the lower or operating level of management.

In today's environment, it is important for an airline to avoid duplication of work structures and improve internal communications where possible. It is also important to create a flexible corporate structure that can expand when necessary and contract if needed without serious harm being done to the business. New start-up carriers have the advantage over existing airlines of being able to tailor a corporate structure that best fits the organization. New corporate structures should provide more authority to individuals at different levels. As authority is delegated, responsibility should be increased with specific positions therefore changing the nature of the typical top-down or silo system to more of a flat organization. *Figure 16.2* visualizes a possible corporate structure suitable for the current aviation environment.

## Fig. 16.2
### The New Corporate Structure

TOP MANAGEMENT

President/C.E.O.

Vice-President

OPERATING MANAGEMENT

Director of Safety

Direction of Operations

Director of Maintenance

Chief Pilot

Chief Inspector

The structure previously discussed will also allow the airline to become more diverse when needed. Diversification is one key to success in the airline business but most airlines are not able to diversify despite having all the resources to do so. The typical pyramid structure is designed so that virtually all decisions for the organization are made among a handful of people. As a result, the talents and skills of others are often not utilized and decisions made are sometimes not in the best interest of the airline but in the best interest of upper management of the board of directors. A less formal organizational structure allows diversification to happen because more skills and talents can be tapped into. Access to such resources will permit the airline to initiate new departments when needed and increase the ability of the company to become involved in businesses outside of the core business.

## Corporate Culture and the New Workforce

Over the next ten to fifteen years, a new generation of workers will emerge resulting in the need for change within many industries including the airline industry. The new generation workforce is referred to as *Generation Y*. Such workers could mean the difference between success and failure and companies must plan now to be successful tomorrow. Much of this planning centers on recruitment and training. The *Generation Y* employee is educated and highly skilled and it is important for management to let the employee know what they do matters. Honesty will go a long way in terms of increased productivity. In other words, the employee wants to be told the truth and they demand that an explanation be given as to why they are being asked to perform a task. Learning the language of the *Generation Y* employee is very important for management and training must be implemented immediately in order to stay ahead of the eight ball. This employee is constantly on the look out for rewarding opportunities and wants to be praised in public. The work place must be a fun environment and the tools must be provided to the employee to do the job successfully. It is time for corporations to revisit their corporate culture and restructure the work environment to increase productivity. As the saying goes, "a happy employee is a productive employee". U.S. based Southwest Airlines™ and Canadian based WestJet™ are perhaps the best examples to follow in terms of creating a positive corporate culture.

In order to cater to the changing workforce, organizations must create conditions that attract the best people. This might mean rethinking the role of the core group within an organization. Building and using a large and diversely skilled talent pool is important as is offering quick and intensive training to increase the employee's value to the company. Training should be results oriented meaning that everyone should be taught career-effectiveness skills. Managers should be taught to manage and employees should be provided with frequent and coaching-style feedback to identify strengths and improve on weaknesses. The *Generation Y* employee is never satisfied with a little knowledge and needs to be given opportunities to pursue continuous training and development. Personal retention plans should be a part of every new employee and the company should offer a resource center for personal growth and development.

## Changing Nature of Human Resources

Aviation organizations should realize that people are the biggest asset. "I am convinced that companies should put their staff first, customers second, and shareholders third," noted Sir Richard Branson, of Virgin Atlantic Airways™. More and more organizations believe in the same philosophy stimulating companies to look toward the future and identify human resource needs.

Airlines must learn to identify human resource needs through the formulation of objectives, policies, and budgets. Strategies should be related to human resource needs temporarily and permanently. Specific jobs should be outlined with specific job descriptions, and only qualified candidates should be recruited to fill positions. Modern recruitment methods include industry contacts, professional recruiters, employment agencies, colleges and trade schools, and various forms of advertising.

In terms of training, employees should be trained specifically in the area for which they were hired. Such training should permit for more advanced functions within the organization and should be able to address social and economic changes that affect the way the organization must operate. All training programs should have some sort of evaluation process to measure performance of the employee and the benefits received by the organization.

There are some barriers or challenges airlines will face in terms of human resources during the course of this century. These include:

1. *Skills.* Many of the skills used by the airline industry are exclusive to aviation. Such skills are costly and time-consuming to acquire. There is a need for constant refinement of regulatory, technological, and market developments. The airline industry is highly cyclical, which leads to overcapacity in human skills and tangible resources.

2. *Need for new skills.* Increasingly competitive environments generate the need for new skills. To be successful in today's airline industry, workers will need specific skills. For example, multilingual, culturally sensitive, and responsive customer-contact will be in demand.

3. *Finding the right staff.* Airlines have realized that finding the right staff is no longer sufficient. The delivery of high-quality service is based on attitudes

and values of employees. For example, much of Southwest Airlines' success is based on a unique corporate culture that promotes positive attitudes.

4. *Labor trends.* Airlines of the future will find it beneficial to use more part-time and fixed-time staff. Charter airlines have been doing this for years, but is a relatively new phenomenon with the schedule airlines. This creates a challenge for mold-acculturated, committed team members.

5. *Multiskilling and flexibility.* The focus of discussion at many airlines is changing from, "Why" to, "How" and to, "In return for what?" Encouragement of productivity growth through multiskilling (the application of multiple skills by one person) and more flexible work practices in highly unionized environments will create a challenge for the airline industry.

6. *Control of labor costs.* Airline passengers are becoming more knowledgeable and demanding, creating a challenge of how to control labor costs without disrupting customer service. There is a strong argument to place greater emphasis on productivity improvement rather than on salary and benefit cuts.

7. *Cross-utilization of human resources.* There will be increased cross-utilization of human resources within global alliances. The challenge is that variables relevant to the attraction, utilization, and motivation of talented employees differ widely between cultural settings. Some unions think that a global labor pool will create a threat to work conditions and job loss.

8. *Making human resource strategies adaptive.* This is the least specific challenge of those introduced, but is it the most significant. Human resource strategies should be as adaptive as corporate and competitive strategies have to be in the face of increasingly complex and turbulent environments.

As indicated, human resource departments are very important to the success or failure of an organization. Paying close attention to the challenges presented will help aviation organizations achieve efficiency and success in the future.

## Summary

Creating a management team for an airline is a difficult and time-consuming process but it is something that must be done correctly from initial stages of development. This chapter discussed the different factors that must be considered when putting together a management team and the reader learned that the right person must be hired for the right job. Different options were presented in terms of how to set up a management team ranging from full-time employees to interim decision makers that could aid in the raising of capital and application for certification. Different organizational structures were also presented and some of the main advantages and disadvantages of each type were discussed. The subject of corporate culture and the need to create a happy and productive environment for employees at all levels was discussed alongside the new generation workforce known as Generation Y. Finally, the changing nature of human resources was brought forward identifying challenges that must be overcome by the airline to maximize future success.

## Key Terms

Top management

Middle management

Lower management

Administrations

Departments

Divisions

Diversification

Corporate culture

Generation Y

## Review Questions

1.  What is the role of the management team and what type of individual should be on board? Has the right person been selected for the right job?

2.  What does each member contribute to the overall success of the organization?

3.  What are the different types of organizational structures that exist and what is the best structure for the airline being created?

4. Has a suitable corporate culture been that will adapt to a changing workforce been designed?

**Web Sites**

http://www.auxillium.com/culture.shtml

http://www.gobenchmarking.com

http://www.google.com/corporate/culture.html

http://www.southwest.com

http://www.westjet.ca

**Recommended Reading**

Holloway, Stephen.   *Straight and Level: Practical Airline Economics.* Burlington, VT, and Aldershot, UK: Ashgate Publishing, 1993.

Morrell, Peter. *Airline Finance.* Brookfield, VT: Ashgate Publishing, 1997.

Shaw, Stephen. *Airline Marketing and Management. 4th Ed..* Burlington, VT: Ashgate Publishing, 1999.

Tretheway, Michael and Oum, Tae.  *Airline Economics: Foundations for Strategy and Planning.* Vancouver, Canada: University of British Columbia, 1992.

Wells, Alexander T., and Wensveen, John G. *Air Transportation: A Management Perspective, 5th ed..* Belmont, CA: Thomson-Brooks/Cole™ , 2004.

**Chapter 17**

## Risk Factors

**Chapter Checklist**

- High barriers to entry

- Additional risk factors

- Exit strategy

- Regulation and liberalization risks

## Introduction

This chapter introduces the reader to the main risks involved in airline operations and provides strategies on how to reduce such risks. Modern airline business plans should incorporate strategies to reduce risks where possible. It is important for this section of the airline business plan to introduce the risk involved with the airline business. Don't hide the risks! This chapter brings forward the need for contingency planning. Additional topics include industry conditions and competition, implementation of growth strategy, operating in international markets, government regulation, number of aircraft, fuel, seasonality and cyclicality, labor relations and availability, dependence on key personnel, reliance on third parties, lack of prior operating history, and lack of public market. One of the main trends occurring in the current global airline industry is the move towards deregulated environments. Therefore, this chapter concentrates on airline operations within a deregulated environment but acknowledges the subject of risk associated with regulated and liberalized environments.

**Total Page Length: Approximately 4 Pages**

## High Barriers to Entry

In regions of the world where deregulation is the norm, one of the expectations was that carriers would have relatively free access to markets because of the mobility of the airlines' chief assets – aircraft. Carriers dominating individual markets would not charge monopolistic fares, according to this theory, because of the ease with which a competitor could enter the market and compete with the incumbent carrier by charging reduced fares. Thus, the mere threat of entry was expected to discipline pricing. Substantial new entries did occur during the early phase of deregulation, but since the mid-1980s, the pace has slowed and the industry has become more concentrated.

Access to many markets has become extremely difficult in recent years because of the difficulty of obtaining terminal space at many hub airports and the risk associated with competing with an airline at one of its hubs. A competitor that wishes to challenge another carrier at its hub faces considerable financial outlays. The cost of providing a competitive level of service at a hub is substantial: expenditures for advertising, personnel, and aircraft operations are crucial during start-up, when the competitor attempts to win business away from the major carrier. The risk of being unable to recover these outlays is the largest single deterrent to entry at hub airports.

It is difficult to compete with a major carrier during start-up because the major carrier has inherent advantages; some result from the scope of its operations, others from marketing. However, there are some exceptions in the United States. Examples of airlines that were able to compete with major air carriers from the start are AirTran Airways™ and JetBlue™. The latter example was the most heavily capitalized airline ever with $130 million in start-up capital. In the year 2004, Richard Branson plans to launch a new airline to compete directly with Jet Blue and will most likely succeed due to high capitalization.

The larger network of the major carrier allows it to increase service at a lower additional cost. In addition, by having an extensive network, the major carrier is more likely to attract passengers, who then form impressions about the quality of service on other routes. Marketing builds on these advantages. Frequent-flier programs make it difficult to lure business travelers away from an incumbent carrier with which they may have already accrued a substantial account balance. If the incumbent has already established preferred-provider relationships with most of the travel agents around the hub, the new entrant faces an additional competitive advantage. Thus, during the months in which a competitor first takes on a major carrier at its hub, the competitor must offer substantial levels of service, which at a minimum include dozens of flights a day. It must also lure frequent fliers away from an incumbent that offers them more opportunities to earn mileage and somehow win over travel agents who have preferred carrier relationships with the incumbent. It should be noted, that in today's aviation environment, passenger loyalty is becoming more difficult to achieve with increased competition.

One alternative that the newer carriers have attempted is to focus on another airport serving the same city. These airports have considerably lower traffic volumes than the major airports that serve those communities, but they have allowed new entrants to develop niche markets. Examples include Midway in Chicago and Stansted in London. Many of these secondary airports offer attractive incentive packages to attract new business for carriers of all sizes.

Airport terminal capacity can also be a barrier to entry for new and existing carriers seeking to enter new markets. Entering a market requires the ability to lease or develop gates, baggage handling, and airport maintenance facilities, and ticketing and passenger waiting areas. Little underused gate capacity and related terminal space is available at major airports in the short term. Over the longer term, it is possible for carriers to enter many markets, but the experience of recent years indicates that such an entry is neither easy nor inexpensive.

Airport operators believe that existing capacity limits are exacerbated at many airports because the incumbent airlines, holding long-term leases with majority-in-interest (MII) clauses or exclusive-use agreements that are able to block airport expansions that would provide more capacity for new entrants. In additions, many airport-airline leases contain clauses that prohibit the airport from charging "additional rates, fees, and charges" and from changing its method of calculating landing fees. The airlines can block expansions with these provisions, but only those that would increase their costs without their consent.

Another barrier to entry has become the presence of dominated hubs. As carriers build the connection banks required to make a hub work, their presence in the local market can become so pervasive as to approach being a monopoly. Airlines use hubs to shield some of their output from competition. As more flights are connected to a hub, the number of passengers available to support additional flights grows. Making the connecting banks work for these flights requires many gates because of the desire to minimize the delay between connections. Also, higher-yield originating passengers help provide the numbers needed to support frequent hub service. Because few airports have excess capacity in the short run (and few have enough local traffic to support more than one extensive network of nonstop service), hubs tend to become dominated by one or two major carriers who use up the existing capacity.

Finally, during the 1980s in the United States, many new entrants were able to begin operations with used or leased aircraft. Many of the older, noisier, Stage II aircraft that are in operation today, had to meet higher Stage III noise criteria by 1999. The new restrictions reduced the supply of aircraft and required carriers to retrofit or re-engine existing Stage II aircraft. Hushkits and re-engine programs were developed for some aircraft. The cost and availability of conversion programs for some of the major aircraft of the fleet were high. In any event, the phase-out of Stage II aircraft has increased the cost of entry to the airline industry by reducing the supply of used aircraft and increasing the cost of operating used aircraft. Today, there is more pressure on operators to use Stage III aircraft as the new Stage IVs are manufactured.

## Additional Risk Factors

Long-term success in the airline industry is difficult and the risk factor section of the airline business plan should acknowledge that management is aware of the risks. The business plan should incorporate contingency planning where necessary in order to reduce the risks involved. Additional risk factors that must be discussed include: industry conditions and competition, implementation of growth strategy, operating in international markets (if appropriate), limited number to start (if a new start-up), rising cost of aircraft fuel, seasonality and cyclicality, labor relations and availability, dependence on key personnel, reliance on third parties, lack of prior operating history, and lack of public market.

Perhaps one of the biggest risks rests in regulatory and legal issues. This section of the airline business plan answers the question, "Will the regulatory authorities let us do this?" Additional questions to be answered are:

- Is it legally feasible?
- Can we legally offer the product/service?
- Is it regulated or non-regulated?
- If regulated, are there any regulatory restrictions?
- Can we surmount these and, if so, is the opportunity potential worth the effort?
- What is the regulatory/government attitude toward this type of product/service?

## Exit Strategy

The business plan must include an exit strategy to give the reader comfort that risk is under control. The exit strategy answers the question, "If we are wrong, if something we hadn't counted on comes up and we need to get out of the business, how do we do it without "losing our shirt" or damaging our reputation?" The exit strategy should describe how the airline will minimize losses (sold and unsold tickets, purchased or leased equipment, established infrastructure, personnel) and describe how possible damage to company and reputation can be minimized. Reputation is very important in the airline business because the industry is small in size despite the global networks. Unfortunately, many people that have failed in the airline industry have been given opportunities to fail again without any consequences. Many airlines go out of business within the first few months of operation and reemerge with similar management under a new name. Developers of an airline business plan should be cautious of this when it comes to selecting the right management to run the company.

## Regulation and Liberalization Risks

This chapter concentrates on risk factors associated with flight operations in a deregulated environment. However, it is important to note that some regions of the world operate in a regulated environment (government-controlled) or a liberalized environment (partial government control). Some of the risks identified in this chapter apply to both environments but additional risks might be incurred depending on what region of the world the airline is to operate in. The business plan should acknowledge local restrictions accordingly. That being said, the business plan should also acknowledge the strengths associated with operating in either a regulated or liberalized environment. Generally, there will be less competition because of increased government restrictions on air carriers. Specific countries within Asia, Latin America, and the Middle East are good examples to research concerning regulated aviation environments. Western Europe (European Union) is a good region of the world to research concerning a liberalized aviation environment.

## Summary

This chapter concentrated on the main risk factors associated with airline operations. When designing the airline business plan, the risk factor section should be based on the type of environment in which the company will be operating. For example, regulated, liberalized, or deregulated. Different regions of the world have different laws and restrictions and the business plan must thoroughly research this. One of the major risk factors discussed included access to markets. New-entrant carriers have difficulty competing against established and major carriers resulting in the implementation of strategies to achieve success. One such strategy includes utilizing a secondary airport facility near a major hub airport resulting in the establishment of a niche market. Aircraft technology is also a major risk factor to overcome. Governments are putting increased restrictions on aircraft operators due to increased environmental regulation forcing new and established airlines to lease or purchase newer equipment. Modern technology is expensive to acquire making it more difficult to raise the funds necessary to operate. Additional risks presented included industry conditions and competition, implementation of growth strategy, operating in international markets, number of aircraft, fuel, seasonality and cyclicality, labor relations, dependence on management, reliance on third parties, lack of prior operating history, and lack of public market. An exit strategy was also discussed comforting the reader that management has a contingency plan to exit the airline business if necessary without damaging reputation.

## Key Terms

Risk factors

Regulation

Liberalization

Deregulation

Exit strategy

## Review Questions

1. What is the definition of risk factor?

2. What are the main similarities and differences between regulated, liberalized, and deregulated aviation environments? What are the main risk factors associated with operating in each of these environments?

3. What strategies does the airline business plan implement to reduce risk?

4. What is meant by the exit strategy? Does the airline business plan incorporate a sufficient exit strategy?

## Web Sites

http://www.faa.gov

http://www.caa.co.uk

http://www.jaa.co.nl

http://www.iata.org

## Recommended Reading

Dempsey, Paul Stephen and Gesell, Laurence. *Airline Management: Strategies for the 21st Century.* Chandler, AZ: Coast Aire Publications, 1997.

Morrell, Peter. *Airline Finance.* Brookfield, VT: Ashgate Publishing, 1997.

Radnoti, George. *Profit Strategies for Air Transportation.* New York, NY: McGraw Hill, 2002.

Shaw, Stephen. *Airline Marketing and Management, 4<sup>th</sup> Ed.*. Burlington, VT: Ashgate Publishing, 1999.

Tretheway, Michael and Oum, Tae. *Airline Economics: Foundations for Strategy and Planning*. Vancouver, Canada: University of British Columbia, 1992.

Wells, Alexander T., and Wensveen, John G. *Air Transportation: A Management Perspective, 5<sup>th</sup> ed.*. Belmont, CA: Thomson-Brooks/Cole™, 2004.

# Chapter 18

## Invitation to Participate – Finding Investment

**Chapter Checklist**

- Today's investor

- Timeline for investment and operations

- Who's going to pay?

- Total amount of investors sought

- Individual investors

- Expected rate of return on investment

- Investment terms

## Introduction

Finding investors for an airline whether a start-up or an existing carrier seeking expansion, is a challenging proposition. This is particularly true in the early twenty-first century as an already risky industry has become even riskier. The failure rate of airlines is high mainly influenced by under capitalization, over expansion, increased operating costs, competition, and business plans that were not complete or flexible enough to accommodate changes in the operating environment. Existing airlines have had to battle with further complications like the terrorist attacks of September 11th, 2001, SARS, and the Iraqi crisis. Between April 2002 and April 2003, U.S. airline stocks decreased in value by approximately 80% per share resulting in the industry requesting up to $13 billion in loan guarantees from the U.S. Government. How is it possible to attract investors to such a risky business? The answer lies in a well-developed business plan. A good airline business plan is an "invitation to participate" because it inspires the reader to become involved with the business concept. That being said, the final element of the airline business plan is the formal invitation to participate. This element of the airline business plan is used mainly for the purpose of seeking investment. This brief chapter introduces the reader to the information demanded by potential investors. Timelines alongside capital requirements are discussed.

## Total Page Length: Approximately 1 Page

## Today's Investor

Many investors today consider the major airlines to be day-trading vehicles because the industry is unstable causing stock values to fluctuate. Investors interested in good balance sheets and earnings tend to avoid such companies because of this risk. However, there is always interest in the airline business because of the great opportunities that exist and as the industry turns from unstable to stable, it is expected that airline stocks will increase in value up to 300% between 2003 and 2005. Proof of the current risk involved in the airline business can be found in the airlines' inability to obtain credit.

When looking at the point-to-point carriers, it seems investors are more attracted to this aspect of the business because risk is reduced compared to the majors. Point-to-point airlines have significantly less operating costs and often operate in markets where there is little to no competition. Because the point-to-point carriers are not dependent on connectivity at hub airports, there is less pressure on increasing demand because such carriers do not have to worry about the excess capacity issue (abundance of seats that remain unsold).

New airlines starting out, particularly low-cost new entrants, have a great advantage over existing airlines because they are able to start with a fresh balance sheet and little debt. They also have the advantage of being able to establish niche markets with strong demand utilizing a simple fleet of aircraft. This is attractive to the investor who tends to view the airline industry as a long-term high-risk gamble.

*Figure 18.1* provides a sample of how the Invitation to Participate might appear within the airline business plan.

## Timeline for Investment and Operations

The Invitation to Participate document is short in length and should not exceed one page. This document is the final element or piece of information supplied in the airline business plan. At a quick glance, a potential investor should be able to see what type of activity is planned during the ramp-up period assuming the business plan is for a start-up. The ramp-up period is usually 12 to 18 months and it refers to the time required to formalize the business leading up to day one of flight operations. The ramp-up period normally commences once the initial capital has been secured. During this time, the airline applies for certification with the appropriate government body, acquires the aircraft, hires the staff, implements required infrastructure, and establishes contractual agreements with the different parties involved in airline operations.

## Who's Going to Pay?

Unless one has a rich uncle, determining who will provide the capital necessary to launch operations will be a challenging process. Not only will many barriers be faced, but timing will be of the essence. Once the airline business plan has been completed, there is a relatively short time span for raising the initial capital. The airline business plan is designed around an opportunity normally in the form of an untapped market. In many cases, such markets are seasonal and it is crucial that operations begin by a specific date as indicated within the airline business plan or the opportunity may pass by and never be available again.

In many cases, the developer(s) of the airline business plan has the knowledge but lacks required funding causing dependency on others to invest into the company. Unfortunately, as capital is raised, it is often the person or group with the money that takes control of the company. If this is the case, it is not uncommon for the vision of the airline to change due to differences of opinion. An excellent business plan can become a poor business plan over night if the wrong people are brought into the picture. When seeking outside investment, be cautious of who becomes involved so that the vision of the airline stays in tact.

Money can be raised through different sources but the common methods are banks, investment houses, and wealthy individuals. Some start-ups have attempted an Initial Public Offering (IPO) trying to raise capital on the stock market. Although this sounds like a good idea in theory, it is not. An airline should not go public until it is operating securely and has had time to establish itself as a successful carrier. Listing on the stock exchange is worthwhile after the airline has been in operation for at least two or three years and has plans to expand. The main purpose for going public is to raise the funding required to initiate expansion plans. It should be noted that once the airline is traded publicly, outside influences will determine the success or failure of the company. For example, management personnel may change and the Board of Directors will have certain demands that did not exist when the airline was operated privately.

Locating a company or an individual specialized in the art of raising capital is perhaps the most common method for turning an idea into a reality. Financial experts will represent the airline and establish meetings with potential investors using an expansive contact list and a "road show" philosophy. Once inside the boardroom, it is up to you and your business plan to sell the concept. Therefore, it is wise to have a well-structured business plan in place prior to discussing any type of financing. Just like the aviation industry, the finance community is small and word quickly gets around if you haven't done your homework. That being said, the same is true when a successful business plan has been presented.

## Total Amount of Investment Sought

This element of the airline business plan should indicate at a quick glance how much the total initial investment is. Potential investors will realize that additional funding will be required in the future but it is wise to show how much is required at specific periods in time. For example, after the ramp-up costs have been finalized, how much capital is required in years 1 to 5 based on the documented growth strategy? The Invitation to Participate should include one figure representing the total amount of capital required in order to launch the airline. It is expected that multiple contributors will provide the requested amount of capital.

## Individual Investors

The airline business plan is designed around a certain philosophy and that philosophy can quickly change as more participants become involved. Therefore, keeping the number of investors to a minimum is preferred over a large number of investors. In other words, the airline stands a greater chance of being successful with fewer participants with large contributions of capital versus a greater number of participants with smaller contributions of capital. Sharing a vision with a small group is much easier to achieve than with a large group. Also, control issues arise with a greater number of participants potentially causing the airline business plan to lose its focus. This element of the business plan should indicate to the reader what the minimal amount of investment is. The advantage to inserting an actual dollar figure helps in determining serious investors from the not so serious ones. Again, time is of the essence and there is no point in wasting time with a potential investor if they do not have the means or interest in contributing what you seek.

## Expected Rate of Return on Investment

One of the most important factors a potential investor seeks when reading a business plan, is how much money will be made on the investment and in what period of time. This is commonly known as the investor's return on investment (ROI). This information must be included in the Invitation to Participate or else the reader's interest will be lost. When it comes down to the basic facts, the investor needs to know how much money is required, how and when it will be spent, when a return is expected, and how much that return will be. If the airline business plan does not satisfy these demands, then the business plan has not met the tasking of acting as a selling tool to raise money.

## Investment Terms

Once a potential investor decides to move forward, the company must discuss what the terms of the investment are. Prior to such a discussion occurring, the current management team should have an idea about what those terms are. It is advisable to consult a finance expert as well as a specialized lawyer prior to establishing any contracts.

Depending on the type of relationship established between the company and the investor, there are a number of factors to discuss in terms of roles and authority, salaries and/or profit sharing issues, short, medium, and long-term goals, as well as foreign ownership rules and regulations, if applicable.

## Figure 18.1
### Invitation to Participate

*X Airways* is poised to apply this business plan towards implementing operations within 12 to 18 months of initial capital investment. As described in this prospectus, *X Airways*, seeks a total initial investment of $X million USD towards company certification under Federal Aviation Administration Part 121, acquisition of aircraft, hiring of staff, and implementation of all required infrastructure and contractual agreements.

Investors are invited to participate in this exciting venture. The ideal participant would be willing to participate at a total of no less than $X million USD, contributed with 10 percent of the investors total contribution to be issued for initial phases of company development and a commitment for the remaining contribution to be issued in readiness for the first year of *X Airways* service operation.

As estimated within the Financial Analysis section of the plan, *X Airways* expects a 10 percent annual return on initial investment beginning quarter 10 of *X Airways* operations. A summary of investment terms will be provided with the support of appropriate underwriting commitments.

Any questions or comments regarding this plan should be addressed to *X Airways, LLC*. (Insert address, telephone/fax contact, and e-mail address).

## Summary

Despite the risk involved with investing in an airline, the number of potential opportunities that exist is great given the right circumstances. A well-developed business plan acts as an invitation to participate because it draws the attention of a potential investor. The final element of the airline business plan is the Invitation to Participate. Although this portion of the business plan is short in length, it is perhaps the most powerful page because it sums up the most important information an investor seeks when it comes to providing capital. The Invitation to Participate tells the reader how much money is required, how and when it will be used, when the return on investment is expected, and how much the return will be.

**Key Terms**

Ramp-up period

Initial Public Offering (IPO)

Return on Investment (ROI)

**Review Questions**

1.  Given the current state of the airline industry, what are the different types of investors that exist and will they be interested in my airline business plan?

2.  What is the purpose of the Invitation to Participate and does it fulfill the potential investors' demands? In other words, is a reasonable time-line provided describing the ramp-up period and how such money will be allocated?

3.  Who is going to invest in the airline business plan and have I considered all the advantages and disadvantages of the different funding sources?

4.  How much money is required to launch operations? When is the expected rate of return? How much will the return on investment be?

5.  Once an investor has agreed to establish a contract, what will the terms of investment be?

**Web Sites**

http://ezinfofind.com/2003/

http://www.504bank.com/VC.asp

http://www.ubs.com

http://www.apig.com

http://www.multexinvestor.com

**Recommended Reading**

Morrell, Peter. *Airline Finance*. Brookfield, VT: Ashgate Publishing, 1997.

Radnoti, George. *Profit Strategies for Air Transportation*. New York, NY: McGraw Hill, 2002.

Shaw, Stephen. *Airline Marketing and Management, 4$^{th}$ Ed.*. Burlington, VT: Ashgate Publishing, 1999.

Tretheway, Michael and Oum, Tae. *Airline Economics: Foundations for Strategy and Planning*. Vancouver, Canada: University of British Columbia, 1992.

Wells, Alexander T., and Wensveen, John G. *Air Transportation: A Management Perspective, 5$^{th}$ ed.*. Belmont, CA: Thomson-Brooks/Cole™, 2004.

# Part Three

# Getting In the Air

# Chapter 19

## Certification

**Chapter Checklist**

- Purpose of certification

- Types of certification

- Timing of certification

- Certification process

## Introduction

Certification is not a topic generally covered in detail within a normal airline business plan. However, it is included in this book in the interests of completeness, and to ensure the most thorough and applicable analysis of issues associated with the airline startup process.

Certification is the point at which regulatory authorities become involved with the business procedures of establishing a new airline. The purpose of certification is to ensure that any and all functions of an airline's operational methods and procedures are certified as safe, as defined by the regulations set forth by government agencies such as the U.S. Federal Aviation Administration (FAA) and the U.K. Civil Aviation Authority (CAA). Each country has an aviation regulation body, and each controls and regulates the safety concerns and, in some cases, economic concerns, of airlines based within its respective country.

Recently, the development of the European Union-based Joint Aviation Authority (JAA), based in The Netherlands, has meant that where once, countries had independent aviation regulation agencies, most European countries now fall under the certification and control of the JAA, through the authority of each individual country's regulation board. This has, in some cases, enabled common certification among most European countries.

This chapter introduces the reader to the certification process. The U.S example will be discussed in detail, walking the reader step-by-step through the five phases of certification including: pre-application, formal application, document compliance, demonstration and inspection, and certification.

## The Purpose of Certification

Aviation certification exists primarily, regardless of country, to ensure that commercial airlines, and all other operators of aircraft, meet minimum safety standards before beginning operations. Once operating, air carriers maintain their certified status under the close eye of the certificating agency.

In order to meet the requirements for certification, most countries require that minimum financial, safety-related, and ownership rules be adhered to, and requirements met. For example, the European Union (E.U.) currently caps foreign ownership on airlines to a maximum of only 49% of total airline equity. The same can be said of many nations. At the time of writing, the United States also imposes foreign ownership restrictions on airlines. A maximum of 25% of the stock (voting) of any airline based in the U.S. may be owned by companies based outside the United States. However, this may change, pending the results of ongoing negotiations between the U.S., the European Union, and other nations.

Within the European Union specifically, there are also ownership and operating laws pertaining to ownership between E.U. member countries. No more than 49% of an airline may be owned by a company based outside of the country in which the airline is based. Such rules apply to airlines wishing to be granted route licenses to operate services outside of the European Economic Area (EEA), essentially defined as the E.U. plus Switzerland, Iceland and all Scandinavian countries. For airlines that are owned fully, or majority, by companies from outside of the specific E.U. country, they are restricted to operations only within the EEA and are not permitted to have route licenses for routes outside of that area.

Financial requirements exist to ensure that the carrier in question has enough financing in place to support operations for up to two years, depending on nation. In some cases, this minimum requirement may include a higher time limit. Some countries do not impose financial constraints on those airlines operating aircraft of a certain size, and some countries do not distinguish between airline sizes or types of operation.

Minimum insurance requirements exist to obtain certification. These requirements cover minimum financial values for hull and liability insurance, business insurance where applicable, and third party damage insurance. As with financial requirements, these can differ depending on the type of certification required.

Certification enables the company to operate under an Air Operator's Certificate (AOC) that governs safety and an Air Carrier Operator's License governing all other minimum requirements. These licenses permit the carrier to fly aircraft of a certain size (or of unrestricted size, depending on the license granted) on a certain collection of routes, where applicable, or within a specified geographical area. Often, however, an Operator's License cannot be acquired until an AOC has already been granted.

## Types of Certification

In the United States and in much of the European Union, once an airline is certified, it is permitted to fly on any route within a given area, with any combination of route frequency and aircraft type. However, what exactly constitutes "any aircraft type" and "route frequency" can vary depending on the country and type of certification.

In the United States, for example, there are essentially two types of operational certificates under which an airline may operate – Type 14 Commercial Flight Regulation (CFR) Federal Aviation Regulations (FAR) Part 135 Commuter/non-scheduled [Charter] operations, and FAR Part 121 Scheduled operations. While both parts of the FARs enable passenger operations, some of the minimums and insurance issues vary between the two types of certificates. Air carrier certification in the U.S. is currently overseen by the Certification Standardization and Evaluation Team (CSET), which standardizes the approach taken.

As stated on the website of the CSET, the certification process is one of "goalposts" – where you must meet certain criteria before continuing to the next stage of certification. For example, it is necessary to obtain a "Certificate of Public Necessity and Convenience" in order to become an approved air carrier and this must be done prior to filing the formal application with the FAA.

FAR Part 121 operations are those under which most major airlines operate. These regulations require that, three months (90 days) before flight departure, a legally-binding schedule be published, by which the airline is legally required to abide.

In addition, FAR Part 121 certification requires the presence and declaration of five specific member positions within the team to start the airline. These positions include a Director of Safety, responsible for interactions with the certification board concerning safety, a Director of Operations, a Director of Maintenance dealing with all maintenance issues, a Chief Pilot, and a Chief Inspector. With proper authorization, however, it is possible to waive the requirements for such positions if the FAA deems it to be appropriate. Such waivers are, however, dependent on the size and complexity of the company structure, and such requests are filed usually at the time of formal application. These positions are discussed in Chapter 16 of this book.

FAR Part 135 operations are those under which some commuter and non-scheduled airlines operate. These regulations were designed largely for air taxi services and charter operations and do not require the airline to operate on a previously-published flight schedule. The result is that, while the operational requirements are less strict with FAR Part 135, some of those requirements – such as the maintenance issues – are actually being brought up to par with Part 121 operations. However, some benefits do exist with Part 135 operations, such as the ability, should it be necessary, to cancel flights without penalty, and remove them from operating schedules, should they exist.

As with FAR Part 121 operations, certification under FAR Part 135 requires certain people within the airline to be declared and specified on the application with certain titles. However, where FAR Part 121 certification requires five members, Part 135 requires a mandatory three including the Director of Operations, Director of Maintenance, and Chief Pilot. For business purposes, however, it is likely advantageous to include other directors in addition, along with financial experts. It is also possible, as with Part 121 operations, to obtain waivers for such necessary positions from the FAA, depending on the complexity of the operation.

In the U.K, meanwhile, under the CAA (and subsequently, the JAA), the distinction for an Air Operator's License is based upon the maximum aircraft size that the carrier may operate. Type A certificates govern all carriers that will operate carriers that operate aircraft over 20 seats, while Type B carriers are restricted to an aircraft of a maximum size of 20 seats.

## The Timing of Certification

Certification should be initiated once the Business Plan has been completed, and once financing has been secured for the venture. The reason that this should not be initiated at an earlier time is that all certification is irrelevant if financing is not available, and it is often the case that certification is not available without specific financing information available.

The projected time between initial application and receiving certification is between 12 and 18 months. Certification takes this long because it is necessary for all application information to be thoroughly processed. Also, the certifying body has many demands placed upon itself because it is responsible for certifying more than one airline at once.

## The Certification Process

In the United States, certification is done, as a general rule, by the Flight Standards District Office (FSDO) nearest to the intended operating base of the airline at the time of its certification application. In other countries, it may be conducted by the parent aviation authority. However, the step-by-step process below concentrates on the certification process for an FAR Part 121 (Scheduled) operator within the United States.

Step-by-Step U.S. Certification Breakdown

The certification process in the United States is divided into five essential parts. The first stage of this is the Pre-application stage, followed by Formal Application, Document Compliance, Demonstration and Inspection, and, finally, Certification.

*Pre-application*

During the pre-application phase, it is necessary to discuss with the FAA and declare exactly what types of aircraft, routes, and facilities intend to be used for operations.  In addition, a meeting will take place between regulatory representatives and company officials, in order to ensure that the company is fully aware of the process into which they must go in order to achieve certification.  It is necessary at this stage to understand completely what is involved with the Formal Application, so that it may be done correctly. The initial stage of application encompasses a scheduled visit to the local FSDO where potential applicants meet as a group to learn more about the certification process. The first meeting is treated very informally.

*Formal Application*

The formal application stage is where, once an understanding has been reached, the company officially outlines its plans.  It is necessary to get all this right the first time, and responsibility for such accuracy falls solely on the shoulders of the applicant company.  The paperwork required at this stage includes a formal application letter stating aims and objectives, and submission of all company manuals by which the company intends to operate. This includes Standard Operating Procedure (SOP) manuals.  The intention here is to ensure that the company meets all operating procedure minimums as set forth by the FAA. All manuals must contain specific information in order to be approved by the FAA. There are a number of consultants within the airline industry that can provide such documentation. In many cases, a consultant will take existing manuals from an FAA approved company and simply tailor the manuals to the likes of the new airline. That being said, the reader should be cautious when it comes to manual production. The management team might have someone on staff that has the ability to produce manuals in-house saving the company substantial money. If the manuals are produced externally, be sure to consult a manual expert with a good reputation and valid references. Such experts often unofficially attend pre-application meetings to introduce him/herself to potential applicants.

*Document Compliance*

Essentially, this is the part of the certification process where the regulatory authority will review and analyze the documents submitted as part of the formal application phase, and will assess the level to which the company follows all legal minimums, as applicable. These minimums will vary, depending on what type of carrier the company intends to operate.

*Demonstration and Inspection*

Some parts of this phase may overlap with parts of the previous phase. In order to ensure compliance with the minimums set forth by the FAA, it may be necessary for the company to either demonstrate some of its procedures – such as those associated with emergency evacuation techniques or cabin crew training – in order to ensure safety. During this phase of certification, the FAA has the right to show up on site unannounced and demand the air carrier to perform flight operations and simulate different types of scenarios. The purpose of such action is to ensure that the company knows how to handle situations when they arise.

*Certification*

Provided that no notably negative issues have arisen from the certification process thus far, or that all issues that have arisen have been rectified, the FAA will at this point issue a certificate that allows the company to operate. The certification provided lays down strict operating rules, depending on the type of carrier, and will ensure that the carrier can meet those requirements.

## Summary

When a new airline is formed, it does not simply go into business and commence operations with passengers and/or cargo on board. Prior to this, the airline must pass a number of steps within the certification process. Air carrier certification is managed by a governmental body responsible for making sure that the airline will meet specific safety and fitness standards. Most countries of the world follow a similar process to the United States, but standards will fluctuate country by country. In the U.S., the certification process is expected to take 12 to 18 months if both the applicant and the FAA work in synch. Delaying responsibilities or neglecting tasks will result in a greater time length and the possibility of missing out on market opportunities. This chapter briefly introduced the reader to certification and the different phases of the certification process. Once the airline business plan is initiated, it is important for the developer(s) of the plan to become familiar with all aspects of certification including those not mentioned in this chapter. It is advisable to research certification issues by directly contacting the issuing body through an official web site or in person. In the U.S., FAA and FSDO officials are more than willing to answer questions and address concerns about certification as it is in the best interest of all parties concerned to operate a successful airline.

## Key Terms

Ownership laws

Operating requirements

Minimum safety requirements

Pre-application phase

Formal Application phase

Document Compliance phase

Demonstration and Inspection phase

Certification phase

## Review Questions

1. When designing the airline business plan, it is important to define what kind of airline the company will be. What different options exist in terms of defining the specific type of air carrier options available?

2. Air carrier certification is completed under the authority of a governmental body. Different countries of the world have different names for such a body. Given the airline's home country, what are the different bodies of government involved with the certification process?

3. In the United States, there are five phases in the air carrier certification process. What are the five phases and what action transpires during each phase?

**Web Sites**

http://www.faa.gov

http://www.jaa.nl

http://www.caa.co.uk

http://cset.faa.gov/

http://www2.faa.gov/avr/afs/fsdo/index.cfm

**Recommended Reading**

Abeyratne, Ruwantissa. *Aviation Trends in the New Millennium.* Burlington, VT: Ashgate Publishing, 2004.

Dempsey, Paul Stephen and Gesell, Laurence. *Airline Management: Strategies for the 21$^{st}$ Century.* Chandler, AZ: Coast Aire Publications, 1997.

Wells, Alexander T., and Wensveen, John G. *Air Transportation: A Management Perspective, 5$^{th}$ ed..* Belmont, CA: Thomson-Brooks/Cole™, 2004.

# Chapter 20

## Achieving Success

**Introduction**

*Wheels Up: Airline Business Plan Development*, is believed to be the first book ever written on how to design and develop an airline business plan using the correct elements. The author's main objective when writing this book was to provide the reader with a comprehensive framework for putting together a solid airline business plan that would ultimately lead to successful operation of an actual airline. It was said at the beginning of this book that there are three types of businesses one should never invest money into due to the amount of risk involved: restaurants, baseball teams, and airlines. It was also mentioned that, regardless of the economic environment, investors are often attracted to airlines because of the huge payback that could result if the company is successful. Admittedly, the airline industry is an extremely risky business but almost any airline can be successful if the right steps are taken from the start. A business plan designed using the contents of this book will provide many advantages for the airline compared to competitors who select the "traditional" methods of building a business plan.

This chapter summarizes the main elements of the airline business plan and encourages the reader to consult previous chapters concerning recommendations on strategies to achieve successful operation of an airline. A step-by-step format is used to refresh the reader on the main topics discussed earlier. The contents of the previous chapters are tied together leaving the reader with many factors to consider before developing an actual airline business plan. No new information is provided. However, the material covered in this chapter is important because all the concepts, strategies, and lessons presented throughout the book are merged into the concluding remarks.

**Part I**

**Step One – Background Research**

Prior to writing the airline business plan, it is important to have a thorough understanding of the global airline industry. Researching historical, modern, and future trends becomes an important aspect of the business plan even if the information is not physically presented. Background information enhances the knowledge base of the developer(s) resulting in better decisions being made as the business plan progresses. *Chapter 1* introduced the reader to valuable information concerning this topic.

## Step Two – Understanding the Purpose of a Business Plan

A business plan acts as a selling tool to raise money and is used as a benchmark or map to indicate how the company is performing. Two types of business plans were identified in this book: the generic business plan and the airline business plan. Although both types of business plans have the same objective, each of them has a different format resulting in either success or failure. A generic business plan consists of elements that can be applied successfully to many types of industries but rarely to the airline business. The elements of the airline business plan are extensive and tailored to the type of industry. Such a business plan is comprehensive because it covers information that would ordinarily be omitted with a generic business plan. More importantly, the elements of the airline business plan permit one key ingredient for achieving success – flexibility. The term flexibility was discussed throughout the book and it was mentioned that flexibility within the airline business plan is important because it allows the airline to adapt to a changing environment when necessary. Traditionally, airlines have not been flexible and able to adapt to changing environments because wrong elements were used in the business plans from the start. Generally speaking, if a generic business plan is used, strategies are often set in stone because flexibility issue has not been incorporated within the main elements. *Chapter 1* and *Chapter 2* introduced the reader to the main purpose of a business plan and identified the differences between a generic business plan and an airline business plan.

## Step Three – Becoming Aware of Common Business Plan Mistakes

Historically, companies in virtually all types of industries make similar mistakes when it comes to the design of a business plan. The airline industry is no exception. Prior to putting pen to paper, the developer(s) must become educated on the common mistakes introduced earlier in this book. Knowledge of common mistakes in advance of designing the business plan will lead to a near perfect airline business plan. Again, the most common mistakes are: failing to capture the reader's interest; inaccuracies, inconsistencies, and lack of objectivity; failure to establish sustainable, competitive advantage; failure to put together a successful management team; and finally, failure to demonstrate revenue growth and profitability. *Chapter 3* introduced the reader to the main issues associated with common business plan mistakes.

**Part II**

**Step Four – Business Introduction (Element One)**

The Business Introduction is the first main element of the airline business plan. This section outlines specific objectives and accomplishments, reveals potential barriers, problems, failures, and risks. It presents the "uniqueness" of the company, market attractiveness, and success factors all in a concise manner. At a quick glance, the reader should be able to have a good idea of exactly what the business concept proposed is all about. *Chapter 6* introduced the reader to the main topics to be included in this aspect of the airline business plan.

**Step Five – The Mission and Vision Statements (Element Two)**

Various factors lead to the success and failure of airlines around the world. Often, although the mission is known, airlines fail to incorporate vision within the airline business plan. As stated previously, the Mission Statement is the portion of the business plan that clearly states the airline's main purpose or aim and provides stability for the airline. The Vision Statement provides flexibility permitting modifications where necessary given changes in the environment in which the airline operates. The Vision Statement is detailed and provides a clear understanding of what the airline's success will look like within a specific time-frame. *Chapter 7* introduced the reader to the necessary content that must be covered when discussing the Mission and Vision Statements.

**Step Six – Strategy (Element Three)**

This element of the airline business plan incorporates the strategic action plan, strategy implementation, growth and expansion strategy, and failures in planning. The strategic action plan is used to guide management with daily decisions impacting both the short-term and long-term operations of the airline. Many airlines have made the mistake of not planning for the future while concentrating too much on the short-term. External forces create rapid changes in the airline industry and airlines find it difficult to succeed because their business and strategic plans are not flexible. The strategic action plan should answer six key questions and once answered, be incorporated into the eleven key elements of the airline business plan. *Chapter 8* introduced the reader to the different factors that must be addressed when discussing strategy.

**Step Seven – Market Opportunity (Element Four)**

Historically, airlines have not done a good job when it comes to market research concerning route networks. Discovering market opportunities is a challenging risk requiring a lot of background research. This old trend of not doing thorough research is changing as the competitive nature of the airline business increases. Research is being spurred by pressure on airlines to reduce costs wherever possible. The main goal of this element of the airline business plan is to provide the reader with a clear view of the market opportunity for the product and/or service. *Chapter 9* discussed the factors requiring research concerning what airports to serve.

**Step Eight – Analysis of Market Demand Levels (Element Five)**

Forecasting techniques are often used for establishing potential demand within specific markets as well as carrying out three important management functions – analysis, planning, and control. The techniques presented in *Chapter 10* can be used to measure quantification using a number of different terms. Forecasting techniques should not be used solely on their own and must be combined with integrated planning techniques. Goals, strategies, alternative courses of action, and a realistic fit with market conditions are necessary for effective results when forecasting.

**Step Eight -- Proposed Route Structure and Schedule (Element Six)**

When discussing the concept of scheduling, it is important to concentrate on the following topics: the scheduling concept, equipment maintenance, flight operations and crew scheduling, schedule planning, and aircraft assignment. Scheduling is a complex process consisting of numerous variables related to all aspects of airline operations. *Chapter 11* introduced the reader to the four main tasks a successful schedule should accomplish: provide adequate service based on passenger demand, provide economic strength for the company in terms of profitability, provide for sales and competitive effectiveness, and provide operational dependability and efficiency.

**Step Nine – Financial Analysis (Element Seven)**

This element of the airline business plan is crucial because it determines success or failure for the airline. *Chapter 12* introduced the reader to the key financial assumptions of the airline business plan including income statement expenses, income statement revenue, operation plan, and summary financial projections. This section of the airline business plan determines the amount of capital required to launch operations and provides valuable information for the investor in terms of revenue, expenses, return on investment, and growth strategies.

### Step Ten – Sales and Promotion Strategy (Element Eight)

A sound business plan does not guarantee success. Market strategies that attract the passengers to the airline must accompany a good business plan. *Chapter 13* introduced the reader to the marketing mix and discussed controllable and uncontrollable marketing-decision variables. It was suggested that a thorough understanding of these variables is necessary before the developer of the airline business plan can establish sales and promotion strategies that will be successful for the type of airline being launched.

### Step Eleven – Aircraft Operating Strategy (Element Nine)

All too often, management makes the wrong decision when it comes to determining what type of aircraft will operate specific routes resulting in high operational costs and inadequate seating to match fluctuations in demand. Although selecting an aircraft is a time consuming process, the extra effort is worth it. *Chapter 14* outlined the major topics that must be addressed in the airline business plan. Such topics include: route and strategy selection, aircraft selection, aircraft type availability, aircraft capacity, aircraft range, fleet and crew commonality, powerplant options, aircraft performance, aircraft efficiency, restrictions and curfews, maintenance issues, as well as aircraft performance and evaluation.

### Step Twelve – Competition and Competitive Response (Element Ten)

This element of the airline business plan identifies strategies the airline must implement to combat current and future competition. *Chapter 15* introduced the reader to the competitive nature of the airline industry and explained why some airlines fail and why others prosper. Although assumptions must be made concerning competitive issues, it is important that assumptions be thoroughly backed by research and forecasting techniques. It is important to develop flexible strategies alongside contingency plans to effectively compete in the world's fiercest industry.

**Step Thirteen – Management and Support Team (Element Eleven)**

Different factors must be considered when putting together a management team to operate a successful airline. It is important to find the right person for the right job. Various options exist in terms of how to set up a management team ranging from full-time employees to interim decision makers that could aid in the raising of capital and application for certification. *Chapter 16* introduced the reader to different types of organizational structures and explained the need for airlines to create a unique corporate culture suitable for the type of organization operated and the type of people employed.

**Step Fourteen – Risk Factors (Element Twelve)**

This element of the airline business plan should be based on the type of environment in which the company will be operating. Different regions of the world have different laws and restrictions and the business plan must thoroughly research this. *Chapter 17* discussed the major risks associated with airline operations including: high barriers to entry, industry conditions and competition, implementation of growth strategy, rising cost of aircraft fuel, seasonality and cyclicality, labor relations and availability, dependence on key personnel, reliance on third parties, lack of prior operating history, and lack of public market. Exit strategy and regulation and liberalization risks were also presented in this chapter.

**Step Fifteen – Invitation to Participate – Finding Investment**

Despite the amount of risk involved with investing in an airline, the number of potential opportunities that exist is great given the right circumstances. *Chapter 18* discussed the need for a well-developed business plan because it acts as an invitation to participate and because it draws the attention of a potential investor. This is the final official element of the airline business plan and it is perhaps the most powerful element of all because it sums up the most important information an investor seeks when it comes to providing capital. This element of the airline business plan tells the reader how much money is required, how and when it will be used, when the return on investment is expected, and how much the return will be.

## Part III

### Step Sixteen – Executive Summary

*Chapter 5* introduced the reader to the Executive Summary and mentioned that this is one of the final steps taken before the business plan is technically considered complete. The Executive Summary is as important as the business plan itself but can only be created once the business plan is ready to be presented in its final form. This document summarizes the business plan without giving away the answers and is referred to as a "teaser package". The Executive Summary has the ability to sell the business idea to a variety of interest groups. This document should be marketed to interested individuals prior to seeing the entire business plan.

### Step Seventeen – The Non-Disclosure Statement

The Non-Disclosure Statement, as discussed in *Chapter 4*, is the document needed to protect the creator of the business plan from having the contents of the plan stolen or manipulated by an outside party. This document should be created once the business plan has been completed and should be signed by the interested party prior to receiving the actual business plan.

### Step Eighteen – Certification

Once the business plan has been completed and all financing requirements have been settled, the certification process begins. *Chapter 19* introduced the reader to the purpose of air-carrier certification and identified the different phases in the certification process. It was suggested that this process not be initiated until the airline is ready to commence operations. This is the final stage of development leading up to the first day of business.

### Summary

There are many lessons to be learned prior to writing an airline business plan. It is important for the developer(s) to thoroughly research the business concept in detail and analyze the successes and failures of other airlines. The airline of today and tomorrow is to be much different than the airline of yesterday. The keys to future success include a solid business plan that encompasses the following: flexibility, diversity, the right management team, the right organizational structure, the right corporate culture, constant training and development for all levels of employees, steady and moderate growth strategies, effective cost cutting strategies that do not jeopardize safety, operation of modern and fuel efficient aircraft, fleet commonality, reasonable capital requirements, and a long-term vision.

The subject matter introduced in this book is extremely valuable to existing airlines, new airlines about to commence operations, entrepreneurs, investors, government agencies, consulting firms, and anyone with a genuine interest in the airline business. The author set out to accomplish a number of objectives realizing that there is a desperate need for a reference source on how to write a solid airline business plan. This reference has been needed for many years and it is hoped that a new foundation has been set for an industry facing great transition over the course of this century. Airlines that use this reference source and follow the established format on how to design an airline business plan, should ultimately increase their chances of survival and success contributing to a stronger industry overall.

One final word…

Despite the risks, instability, and stresses associated with the airline business, there is no other industry on earth that stimulates such excitement, hope, opportunity, and potential prosperity. The sun is always shining above the clouds regardless of the state of the industry or the world surrounding it. Every day is a good day in the airline business!

## Key Terms

Each chapter in this book concluded with a list of key terms that should be identified prior to developing the airline business plan. An understanding of such terminology will help the developer(s) of the business plan in terms of identifying the correct information to be presented in the final document.

## Review Questions

Each chapter in this book concluded with a list of key questions that should be answered as the airline business plan is developed. Although there are many more questions that will be posed and answered throughout the development of the business plan, the author attempted to make the reader aware of the most important ones.

**Web Sites**

Each chapter of this book concluded with a list of reputable web sites providing additional information on the subject matter presented. When consulting web sites for information or advice, it is advisable that the reader seeks mainstream sources that are widely accepted. Although this does not apply to all, be cautious of private corporations on the Web, unless you are familiar with their business. Government, airline, airport, and aircraft manufacturing sites are highly recommended for consultation.

**Recommended Reading**

Each chapter of this book concluded with a list of recommended readings. Where possible, the author listed modern mainstream references held in a high regard by the airline industry as well as academia.

PowerPoint™ Presentation

For readers interested in using the contents of this book in a visual format, a complimentary PowerPoint™ presentation covering each chapter is available at *Aviation Online: The Brooks/Cole™ Aviation Resource Center* (http://aviation.brookscole.com).

Sample Airline Business Plan

As mentioned at the start of this book, a sample airline business plan containing the main elements of an airline business plan is available. The specific business concept created is strictly hypothetical and is based on a fictitious airline called, *Utopia*. The airline is based in the United States and utilizes a fleet of jet aircraft to transport passengers to domestic and international markets. This is a good reference for visual format of the business plan but more importantly, the sample business plan does an excellent job of using Excel™ spreadsheets to breakdown all the financial details of running an airline. Such information includes: expenses, revenues, and growth figures. The author highly recommends consulting this document alongside the book. The sample business plan is available on a CD-ROM and can be obtained directly from the author for a nominal fee. Contact details were provided at the start of the book.